Feminist Technosciences

Rebecca Herzig and Banu Subramaniam, Series Editors

Underflows

**QUEER TRANS ECOLOGIES
AND RIVER JUSTICE**

CLEO WÖLFLE HAZARD

UNIVERSITY OF WASHINGTON PRESS
Seattle

Copyright © 2022 by the University of Washington Press

Composed in Chaparral, typeface designed by Carol Twombly

26 25 24 23 22   5 4 3 2 1

Printed and bound in the United States of America

All rights reserved. No part of this publication may be reproduced or transmitted in any form or by any means, electronic or mechanical, including photocopy, recording, or any information storage or retrieval system, without permission in writing from the publisher.

UNIVERSITY OF WASHINGTON PRESS
uwapress.uw.edu

LIBRARY OF CONGRESS CATALOGING-IN-PUBLICATION DATA
Names: Wölfle Hazard, Cleo, author.
Title: Underflows : queer trans ecologies and river justice / Cleo Wölfle Hazard.
Description: Seattle : University of Washington Press, [2022] | Series: Feminist Technosciences / Rebecca Herzig and Banu Subramaniam, series editors | Includes bibliographical references and index.
Identifiers: LCCN 2021021848 (print) | LCCN 2021021849 (ebook) |
  ISBN 9780295749747 (Hardcover : acid-free paper) | ISBN 9780295749754 (Paperback : acid-free paper) | ISBN 9780295749761 (eBook)
Subjects: LCSH: Rivers—Research. | Environmentalism. | Ethnoscience. | Ecofeminism. | Queer theory. | Sexual minorities.
Classification: LCC GB1201.7 .W63 2022 (print) | LCC GB1201.7 (ebook) |
  DDC 304.2082—dc23/eng/20211119
LC record available at https://lccn.loc.gov/2021021848
LC ebook record available at https://lccn.loc.gov/2021021849

♾ This paper meets the requirements of ANSI/NISO Z39.48-1992 (Permanence of Paper).

For Bo Brown

Contents

Acknowledgments ix
Prologue: Ways to Live Together along Rivers xiii

Introduction: Water's Invitation 3

UNDERFLOW 1: Hyporheic 36

1 Thinking with Salmon about Water: More-than-Human Commons and Hyporheic Imaginaries 40

UNDERFLOW 2: Being in the Field 78

2 Queer × Trans × Feminist Ecology: Toward a Field Praxis 80

UNDERFLOW 3: Trans Thought as Latent to Manifest Destiny Logics 103

3 The Watershed Body: Trans and Queer Moves in Beaver Collaboration 106

UNDERFLOW 4: Making Queer Kin and the Queer Field 142

4 Unchartable Grief: Scientists Grapple with Extinction Politics 146

UNDERFLOW 5: Affects and Ecopoetics Practice 169

5 With and for the Multitude: Cruising a Waterfront with José Esteban Muñoz 174

UNDERFLOW 6: Field Writing with the Brown Commons 207

Epilogue: Upwelling 210
Notes 227
Bibliography 253
Index 273

Acknowledgments

This book is a collective work in so many ways, and many who have shaped its meandering path appear in its pages.

I extend deepest respect and admiration to the land, streams, fishes, herps, plants, microbes, and humans with whom I've worked. This book describes long-standing collaborations in Salmon Creek, the Scott Valley, and the Mid-Klamath and forays to the Eel, Methow, Yakima, and Los Angeles Rivers, as well as an emerging collaboration on the Duwamish. I acknowledge and honor the Indigenous peoples who care and fight for these lands and waters, and I thank the Karuk, Quartz Valley, Stillaguamish, and Duwamish Tribes for their generous invitations to work together to protect these streams.

On Salmon Creek, many residents invited me into their homes and showed me their springs. The Salmon Creek Watershed Council and the Gold Ridge Resource Conservation District welcomed me into their research collaboration and provided logistical assistance during fieldwork. Lauren Hammack, Brian Cluer, Michael Fawcett, Sierra Cantor, John Green, Noel Bouck, Kathleen Kraft, Diane Masura, David Shatkin, Hazel Flett, and Erna Andre were especially critical to my work; Brock Dolman and Kate Lundquist's beaver puns brought laughs and inspiration.

At the Scott River Watershed Council, I thank Betsy Stapleton, Michael Stapleton, Charna Gilmore, and Erich Yokel for collaboration and for mentoring my students in fieldwork and rural relations.

At the Quartz Valley Indian Community, Crystal Robinson shared groundwater data and water politics insights and generously reviewed multiple drafts of chapter 1.

In the Mid-Klamath, Lisa Morehead-Hillman and Leaf Hillman are mentors and inspirations—thank you for welcoming me and my family down to the river. Stormy Saachs and Stefan Dosch shared food, stories, and music. At the Karuk Tribe Department of Natural Resources, I thank Bill Tripp, Kenneth "Binks" Brink, Toz Soto, Mike Polmateer, Frank Tripp, Kathy McCovey, Vikki Preston, Earl Carley Whitecrane, Shawn Bourque, Heather Rickard, and Aja Conrad.

At UC Berkeley, Isha Ray and Stephanie Carlson were incredible cochairs and remain crucial mentors in how to run a feminist lab group; the ERG Water Lab provided critical encouragement at early stages. Laurel Larsen, Mary Power, and Jeff Romm provided insights that shaped my analysis of Scott Valley and Salmon Creek groundwater. Guillermo Douglas-Jaimes, Daniel Sarna-Wojcicki, Sibyl Diver, Kristina Cervantes-Yoshida, Jason Hwan, Kauaoa Fraiola, Suzanne Kelson, Danielle Christianson, Michael Bogan, and Pablo Rodriguez-Lozano provided field support, camaraderie, and solidarity during graduate school. I wrote the first sketches chapter 1 in Kim TallBear's seminar Indigenous, Feminist, and Postcolonial Perspectives on the Environment; that introduction to feminist STS, and reading together across queer and Indigenous relationalities for my quals laid the groundwork for this thought. Ongoing work with Dan S-W informs chapter 1 and the epilogue.

The Water Underground provided a counterpoint to academic study, in performance, in art making, and in the streets: Annie Danger, July Hazard, J. Dellecave, Sarolta Jane Vay, Qilo Matzen, SofT Zulah, Rori Rohlfs, Sarah Kennedy, José Navarette, Ezra Berkeley Nepon, and all the other performers, cast, and crew at the *Gold Fish* performances and filming. J. Dellecave has been a crucial partner in performance and theory interventions at the intersection of ecology and performance, and her multiple readings of chapters 3 and 4 sharpened my understanding of performance studies moves.

At Santa Cruz, conversations with Karen Barad deepened my analysis of what feminist science could be and provided close readings of chapters 1 and 3. The Science and Justice Center fellows and staff welcomed me into lively conversations among scientists and feminist science studies scholars; Kristina Lyons, Jenny Reardon, Donna Haraway, Anna Tsing, Neel Ahuja, Lindsey Dillon, Paloma Medina, Lani Hana, and Sara Mameni gave generous responses to early drafts. Marcia Ochoa's genderqueer solidarity and interest in my reading of Muñoz and the brown commons were critical in shaping my approach while at UCSC, and in weaving queer-of-color critique more deeply into the book.

At UW Seattle, I thank students in my FRESH Water Relations Lab and in the Ecopoetics along Shorelines and Critical and Imaginative Restoration Ecologies seminars and the Ecocultural Restoration and Salmon Science in the Klamath Basin field course. I thank my colleagues in SMEA for the support to develop these unconventional courses and the Honors Program for providing crucial financial and administrative support for the Ecopoetics along Shorelines series. Among colleagues, Maria Elena Garcia, Tony Lucero, Phillip Thurtle, Jean Dennison, Chris Teuton, Josh Reid, Megan Ybarra, Caitlin Palo, Eli Wheat, Kathleen Woodward, Jason Groves, Jesse Oak Taylor, Rich Watts, Chadwick Allen, and Chandan Reddy provided intellectual and moral support. The Feminist STS Writing Group, the Indigenous Studies Writing Group, the Geography Writing Group, and the Trans Futures Writing Group all provided camaraderie and accountability for regular writing.

The FiSTS network has been a wonderful intellectual community; thanks especially to Banu Subramaniam and Angie Willey for the Science out of Feminist Theory special issue and Laura Foster, Deboleena Roy, Harlan Weaver, and Ashton Wessner for ongoing conversations on queer and ecological field practice.

I presented pieces of this book in early form at the UC Berkeley Geography colloquium, the UC San Diego Anthropology colloquium, the UC Santa Cruz Feminist Studies colloquium and Science and Justice Training Center, and the UW Geography colloquium and Program on the Environment Rabinowitz Lecture, where critical feedback helped sharpen the argument. A workshop at the UC Berkeley Queer Ecologies | Feminist Biologies working group was especially generative for chapter 5.

As I was beginning to pull together queer and ecological thought, two extra-academic reading groups showed me the best kind of undisciplined thinking—my deep gratitude for this undercommons and ongoing conversation to Ammi Keller, Heidi Lypps, Brian Peterson, Sarolta Jane Vay, Lindsey Dillon, and the occasional drop-ins. Also in the extra-academic realm, H. Wheels Darling read drafts of chapters, and Ammi Keller worked some literary magic on the beavers.

For being awesome archivists and digging up old performance stills, I thank Ezra Berkeley Nepon, Enid Baxter-Blader, and Qilo Matzen. Many thanks to Michael Bogan, Stephanie Carlson, July Hazard, and Erica Rubenstein.

I am grateful for financial support from the UC President's Postdoctoral Fellowship and the Simpson Center's Society of Scholars Fellowship and for a visiting position at the IAS-STS in Graz, Austria, for crucial time to write. An ADVANCE Transition Fund grant served in lieu of parental leave, for which I thank the anonymous donor. A Doctoral Dissertation Research Improvement Grant and a Graduate Research Fellowship from the National Science Foundation supported my field ecology and hydrology work on Salmon Creek. Grants from EarthLab, the UW Royalty Research Fund, and the Mellon Foundation Just Futures Initiative supported my Mid-Klamath river collaborations; a Northwest Climate Adaptation Science Center fellowship to Jenny Liou supported my work with the Quartz Valley Indian Community. The Whiteley Center and the Neysayer Writer's Retreat allowed me quiet time to write.

July, I drove so many backroads down to the river with you. These stones, your hands on the wheel of the pickup and the van, the poems you wrote there, they are latencies in this work, and traces of journeys yet to come.

Prologue

Ways to Live Together along Rivers

In the Scott Valley in Northern California, tailings stretch as far as the eye can see. The tailings aren't orange, aren't acidic, are just the riverbed itself, churned up, inverted, and dragged out into massive linear mounds, as if a giant ran a rake through a massive Zen garden. Even at ten a.m., the heat beats down on the treeless landscape. A rough road runs along the top of one mound; my friend and sometimes-lover Wheels navigates her Prius expertly among the boulders. To either side, groundwater surfaces in green-blue ponds, clear, deep, and looking very cold. A few solar panels power a submerged antenna that records when tagged fish swim back and forth.

I walk down the hill, toward voices coming from a thicket of willows. Sugar Creek comes into view, and on its banks, the signs of a fish survey—a truck bed full of wetsuits and waders, nets piled by the water, and dozens of buckets with aquarium bubblers hooked over their rims. Arrayed around a table, a motley crew of workers. The Karuk Fisheries crew is in T-shirts and swim trunks, three white women are in waders, and rancher high school kid wears jeans. The crew puts me on the measuring station; I cup my hand gently around a scant handful of slippery young salmon called parr. They are sedated with carbon dioxide released by Alka-Seltzer tabs we drop into the buckets, and they only flop a little on the measuring board and in the plastic bin on the scale. I call out the lengths and weights to Jennifer, a biologist retired from the US Fish and Wildlife Service, then pass the fish to Kenneth "Binks" Brink of the Karuk Tribe's Fisheries Department. He deftly injects tiny PIT tags—microchips the antennas read—into salmon bellies, talking a mile a minute about fish weight, length, and politics.[1] I'm surprised to see the Karuk crew here and say so to Binks.

"I've never been here before," he says. "Charna from the Watershed Council invited us to do the fish handling. I don't come to the Scott Valley unless I'm invited. It's all private property up here; Karuk aren't welcome." To Jennifer, he throws around biologist jargon on life-history models and survival at life stage, comparing fish counts from different sites along the Klamath River and its tributaries. He takes the crew to catch more salmon fry with two-pole seines; the poles are lightweight, made of peeled fir saplings. Jennifer's dog follows them, scaring the fish into the nets. Before she retired, she and Binks worked together on the Mid-Klamath; she helped push federal funding through for the Tribe's off-channel rearing ponds. Back at the tagging table with a new bucket of fish, Charna Gilmore, the Watershed Council director, draws out quiet Clayton, the young Karuk Fisheries technician, about his kids and talks to the high school kid about football and hunting; that gets him and Binks talking about deer hunting spots.

As we're carrying the last buckets of revived parr to release, Binks tells me to grab a snorkel. The deep pool stretches far upstream between tailings walls, backed up behind a woven pole-and-brush structure the Watershed Council built to mimic a beaver dam. When they built this "beaver dam analogue," Charna tells me, the creek barely trickled through the bare creekbed. Now, two years later, willows and alder thickets all but hide the banks, and cattails and rushes grow in the shallows. I put on the snorkel and mask and slip into the icy water. Young coho and Chinook salmon and steelhead trout school around piled-up logs, darting and flickering in the afternoon light. Along the brush dam, more fat coho feed and swim in cold, cold water.

I have special affinity for coho and steelhead, who spend their first summer of life in little streams that flow through the valleys, like the Scott, that settler-colonial society has so transformed. The Scott River's settler impacts—mining damage, irrigated alfalfa, and long reaches that go dry—are familiar patterns up and down western US salmon streams. All six species of Pacific salmonids, which include salmon and the closely related steelhead trout, are affected by centuries of wanton resource extraction and disregard for rivers' dynamic energies. To protect these precarious remnants of once-prolific runs is to wade into the thick of politics and ideologies of "beneficial use" and "public trust."[2]

Current water management debates in US West rivers play out against the uneven spatial distributions and temporal patterns of wet and dry—distributions that are now sedimented into dams, levees, and governance structures. Frontier boosters and dam engineers saw the region's unpredictable precipitation, dry summers, and megafloods as unruly and unproductive. They designed physical and regulatory infrastructures to control water's spatial and temporal unruliness. A century of dam and levee building made wet land dry and dry land wet. In California, levees, built by hand by Chinese railroad workers beginning in the 1870s, then shored up with steam-powered dredges and electric pumps through World War I, dried up marshy islands in the Sacramento–San Joaquin Delta. Drainage projects lowered water tables in major valley wetlands from the Klamath Basin in the north to the Tulare Lake in the south. To avoid reflooding the land, engineers built thousands of dams in the mountains, drowning river valleys, displacing communities, and regulating river flows to produce hydropower and irrigation. By the 1930s, populations of salmon and other migratory species that could not swim past the dams crashed. Such projects exemplify the Manifest Destiny rhetoric of settler exceptionalism, which aims for control of nature, rather than interdependence among species.

Back on Sugar Creek, Charna brings the last bucket of fish down to the stream to cool off. My research partner Dan and my lover July show up after a long, hot drive from Oakland; they run into Wheels and walk down together. We all wade into the water—chest deep and cold—and follow Charna up the channel, carrying buckets of fish to release upstream. Bulrushes and floating plants are slick with diatoms and crusted with aquatic insects. Thousands of fish hide amid plants and in the shallows, feeding on drifting bugs. Ripe blackberries dangle over the water; we eat a few. Charna points out where beavers have dug dens in the banks, chewed sticks, piled up mud on another human-built beaver dam analogue. Nightly, Charna spies on the beavers' labors with motion-activated wildlife cameras. "This guy decided our dam wasn't good enough," she says. "He was working on it a few hours a night, and then, bam, he stayed up working steadily from midnight to six a.m. and plugged all the holes. That raised the water level upstream two feet. Then later, he decided he didn't like it and chewed a hole in it and let all the water out."

We follow a beaver-dug canal into the maze of tailings, then scramble up the cobble slope. After shivering in the deep shade, we're blasted by sun and radiating heat. At the crest, we see the confluence. Below the brush dams, Sugar Creek sinks quickly into the dry bed. In the far distance, sprinklers pivot over alfalfa fields lined by brimming irrigation ditches. Several miles downstream, the stream surfaces again; a few hardy coho wait out the rain's return. The tailings play tricks with perspective, in space but also in time. Gold dredges operated here for a few decades, in a 1930s revival of frontier gold rush fantasies. Dredge tailings obliterated the floodplain's tangled braid of shifting channels and rerouted underflows that kept the river flowing through the summer. These landforms will likely persist for millennia, only gradually softening in once-in-a-century floods. Nonetheless, salmon, birds, and beavers—among other creatures—make their lives here and show unlikely coalitions of humans how to work with the river in a different way.

Think of for a moment of floodplains as amphibious, brought into being and constantly re-created and transformed by interactions among water, sediment, and numerous species (including beavers, willows, cottonwoods, and humans). Beavers cocreate iconic, amphibious landscapes/waterscapes that remix patterns of wet and dry and support floodplain ecosystems in which a multitude of human and nonhuman watershed inhabitants can also live "amphibiously."[3] Large dams and water works have had the opposite aim of beavers': to separate water from land and river channels from their floodplains, to extirpate marshy indeterminacy and replace it with "productive" irrigated farmland and "efficient" channel networks for flood control. Yet for whom are such infrastructures and logics efficient and productive?

Many of the ranchers who have rights to divert the river's surface and underflows to their fields consider property and regulatory regimes to be as permanent as the tailings. The Karuk Tribe and Quartz Valley Indian Community see both as ready for transformation, not back to a pre-European settlement state, but into a new regime that allows salmon to return abundantly. With the Watershed Council and numerous other partners, the Karuk Tribe pursues alliances through fieldwork reciprocity within and outside of their ancestral territory. They and other Klamath River tribes work on behalf of salmon that swim through their territories to spawn here, in Scott Valley, which holds the best coho habitat in the entire Klamath Basin. Their multifaceted natural resource program asserts territorial sovereignty across lands currently held by the Forest Service and exercises treaty rights to comanage fisheries beyond those territories. From these tribes' perspectives, property and water regimes must and will change so that water can flow across floodplains and percolate down, later upwelling into pools where coho and steelhead fry grow fat.

Native peoples in the Pacific West engineered river mouths to let salmon upstream and modified watershed hydrology by strategically burning the uplands and riparian corridors. Such engineering profoundly transformed river ecologies to favor culturally important species like salmon, willows, and other fiber plants that grow in seeps and along sandbars. Settler-colonial water engineers in the late nineteenth and twentieth centuries misrecognized these Native-engineered swamps, marshes, and floodplains as wastelands. Most were blind to, or simply didn't value, the promiscuous ecological productivity of amphibious landscapes. Recent collaborative experiments with beavers, such as the Scott River Watershed Council's at the Sugar-Scott confluence, open up unorthodox ways of understanding and managing rivers in agricultural landscapes: as productive not for capitalist extraction but for fishes, birds, and riparian forests. Such politics inhere in beautiful and troubled rivers like the Scott and the Klamath. These rivers' ceaseless call-and-response of flood and low flows beckon salmon into an ever-changing improvisation of outmigration and return. This book invites river workers and those who love rivers to attune to their unruly and hidden flows and enter into this always-improvisatory work of transfiguring.

Since that hot day in 2016 by Sugar Creek, informal collaborations have grown into formal partnerships. I invited the Scott River Watershed Council, the Quartz Valley Indian Community, the Karuk Tribe, and the Mid-Klamath Watershed Council to collaborate on a field course. The twelve students—some Karuk, some from Quartz Valley, and some settler university students from the University of Washington—learned from these river experts as they surveyed fish, water quality, and habitat. Their reports on conditions in four Mid-Klamath tributaries contributed to the partners' diverse projects that reconnect rivers, floodplains, and groundwater. Upstream from the Scott River's confluence with the Klamath, plans to remove four massive hydroelectric dams are advancing, through public meetings, hauling contracts, plans to monitor how salmon respond, and protests at federal government delays. In Karuk and Shasta ancestral territory in the Mid-Klamath rivers, I didn't know whether my queer, trans, and feminist practices would find welcome or take root into ongoing collaborations. But they did, first with Charna and the Watershed Council on a policy analysis of restoration options and their climate impact, then with the Quartz Valley Indian Community on fish surveys and participatory water modeling. Most recently, Lisa Morehead-Hillman, a Karuk basket weaver and educator, and Leaf Hillman, a Karuk ceremonial leader and cultural practitioner, proposed a project that blends Indigenous storytelling and land-care methods, ecopoetics field writing, and spatial and hydrological analysis. We are deploying these methods within Karuk protocol and a queer praxis of articulating a "river model of justice."

While playing with the symbolic resonance of underflows and other watery words, this book refuses to reduce water to a metaphor even as it recognizes the power trans/figurings that water lets one do. In the parlance of hydrology, water moves through the gravel or sand of the *hyporheic zone* (from the Greek *hypo*, below, and *rheos*, flow), as well as flowing visibly over the surface. *Underflows* draws on this scientific sense of unseen flows to theorize from emerging hydro-ecological thinking on dynamism, from my own scientific research into salmon survival in California streams, and from performative interventions that inject queer and trans brilliance into science and policy realms. In recounting and theorizing from these encounters, I hope to create an opening for you, reader, and for others in your chosen research families to think together across difference, creating spaces of protection, reflection, and mutual recognition that will strengthen movements for ecocultural resurgence.

Underflows

Introduction

Water's Invitation

Rivers matter for queer trans life. Rivers have long been places of encounter for those outside of straight settler propriety; they are brown in the way that José Esteban Muñoz describes, "the way in which they suffer and strive together, but also the commonality of their ability to flourish under duress and pressure . . . they smolder with life and persistence."[1] Queer-trans people make joyous moves of solidarity when we swim in rivers and splash along other shorelines—to wear what we like to the beach, to see and support one another, to be admired in our bodies, without apology.[2] Such everyday acts of life, love, and dreaming big transgress settler norms of sexuality and family and require uprising and political transformation. During a pandemic year, as police violence and federal inaction amplified suffering in Black and Brown communities, Black trans matriarch Ceyenne Taylor shouted "We want to live!" from the stage at the 2020 March for Black Trans Lives. As Taylor told the ten thousand supporters gathered in Brooklyn, for trans and queer people, especially those most marginalized by whiteness and settler colonialism, "society has neither given us our rewards, nor dealt with our deaths in a way where we felt held."[3] Rivers can become spaces to practice these political transformations and transgressions in the everyday.

In these pages, I trace contact points in queer-trans strategy and Indigenous and environmental justice movements and propose that deepening their alliance can push river science and policy to center marginalized peoples' knowledge, relations, and management. By centering queer-of-color and trans critique in environmental politics, I aim to show how Indigenous, Black, Brown, and other radical traditions—in which queer, trans, and Two-Spirit people have always played key roles—are crucial to meeting the intersecting challenges of climate change, settler-colonial legacies of environmental harm, and gendered violence.[4] We settlers have long evaded responsibility for how policies benefit us and harm Native, Black, and immigrant communities. Uniting trans and queer-of-color theory with environmental justice thought has enabled me to develop cross-cultural collaborations that mobilize conceptual underflows: subjugated narratives in science, water politics, and environmental policy. By reading these collaborations as queer trans feminist interventions into river science, I propose ways to transfigure conceptual surface flows: water and salmon governance and watershed management writ large.

Like queer and trans peoples' relations with rivers, our relations with sciences are multiple. *Queer and trans people* refers to the individuals who make up the collective from within which I'm writing, and *queer trans* (without a comma) refers to that collective, which celebrates genderfuck and trans brilliance and includes queers who don't identify as trans but embrace multiplicity. I use *queer trans* to make space for the cis queer and trans nonqueer people within this collective. With regard to science, a growing community is coalescing around the term *queer science*. Many, but not all, of the trans scientists I know also identify as queer and rally around the *queer science* banner. When I specify *trans sciences*, I'm flagging the difference our embodied experience of transition makes in our practices of science. In the title of chapter 2, I use notation from statistical models to interrogate how a practice of river science that is queer, trans, and feminist works and matters. In statistics, the × indicates an interaction between response variables, as when the Mermaid Parade and good weather increase queer trans presence at Coney Island, more than either factor alone.

In environmental politics, I focus on ecological science and the relations among species and earth processes that are this science's domain. I demonstrate how queer trans thought and life can influence ecology (the science) and ecologies (the multispecies worlds). Theorizing from queer trans river relations across fields and research modes, I show, first, that queer and trans people's experiences of grieving premature death—from police and intimate violence, AIDS, suicide, and exclusion from health-care and economic opportunity—can stoke collective action on extinction and ecological repair. Second, I show that trans people's experiences of transfiguring our bodies and social relations model a new way for cis straight people to embrace dynamism and unpredictability, which rivers and their ecosystems enact unceasingly, increasing the adaptive power of restoration ecology.[5]

In restoration ecology, river scientists have paid most attention to ever-varying processes of flow and sediment movement as they effect ecological processes and are influenced by living beings. Species, responding to elementals like fire and flood, make the rivers that queers and trans people love to lounge by and swim in. Trans thought figures deep uncertainty and revolutionary change as a space of possibility, opening, and excitement for the world's becoming. Trans thought thus offers tools to reframe mainstream environmental dissociation from climate change futures.[6] Dystopian narratives too often breed inaction and dissociation among white settlers, who have long avoided responsibility for the dystopian presents our policies have created in Native, Black, and immigrant communities.[7] This book's core argument is that in order to become more adept at centering justice, sovereignty, and adaptability, river sciences should engage with queer and trans theory. I write, in part, for other river scientists, demonstrating by telling stories from my life as a settler queer and trans river scientist—a life that takes place in the presence of love and grief for rivers and fishes.

To trace this line of argument, which is simultaneously queer and trans, I cast my mind into memories of water in its many habits, while sheltering in place in my Seattle home in a winter too cold for naked swimming. The river, any river, invites looking, dipping a toe, diving in, and pulling one's body through bright currents to the dim center of its whorls. Water led me here, to this thought about science and politics in an age of extinction, uprising, and rapid climate change. Water rushed down concrete channels of my first river, the Los Angeles River turned storm drain, but also made cracks in the pavement and seeped down into the vast gravel-and-sand aquifer beneath. One place along that river, at the mouth of the San Fernando Valley, artesian flows surged up over a sill of basalt to heave up pieces of that engineered channel's concrete bed, again and again, until the Army Corps of Engineers gave up paving it. That river often surged to floods that were queer in their excess. These floods no longer recharged alluvial aquifers and fed willowy *ciénegas*, but they filled its concrete channel and threatened low-income neighborhoods with toxic inundation.[8]

Water's examples and invitations made kin of me and water. In reciprocating this invitation, I have made queer trans thought about water's relations along rivers that flow across what's called (for now) the western edge of the United States. I began thinking these thoughts after dropping out of college, during an apprenticeship in arid land rehabilitation. Doing environmental justice education in the San Francisco Bay Area and solidarity work at Black Mesa on the Navajo Nation, I became obsessed with dams as settler-colonial projects and schemed local, sustainable alternatives. In the coedited anthology *Dam Nation: Dispatches from the Water Underground*, I argued that architects of western US settlement obsessed over taming rivers but that the results of their engineering and control schemes are anything but docile.[9] Science directs and arbitrates settler-colonial water management, and I wanted to understand it firsthand.

This book traces the tacking route I took through the sciences, heading first for geology, then social science, then back toward natural science in ecology and hydrology, then toward feminist science studies inflected with queer tactics and a commitment to support Native science and action for ecocultural restoration. Native societies in Pacific salmon territories have long mixed labor with river flows.[10] Some engineered fish passage through sandbars and surveyed canoe routes around obstacles.[11] Others modified watershed hydrology through cultural fire and cultivation.[12] Through government-to-government negotiations, Native nations coproduced robust and genetically diverse salmon populations by crafting protocols for monitoring and harvest.[13] Such Native practices continue into the present, innovating science and engineering with newer tools like bulldozers and statistical analyses.[14]

I present results from straight social and natural science research, interbedded with analyses of queer affects and trans embodied practices that suffused and exceeded the academic and policy products of those investigations. This approach builds theory in the overlay of environmental politics, feminist science studies, Indigenous studies, and queer and trans studies by exploring river cultures, sciences, and politics at five sites of water conflict and river restoration. Grounded in scientific and ethnographic data that I collected as a participant in salmon, beaver, and floodplain recovery projects and in queer trans feminist reflexive writing on performance and art-science work, this underflows approach centers collective practice and relationality within river and ecological governance. As I and water will show, this approach can transfigure riverine relations by highlighting scientists' feelings, solidarities, affects, political frictions, and possibilities: ways of doing science that disrupt settler colonialism's everyday logics.

Sensing Surface Flows and Underflows

When I go to a river—to the Klamath, the Stillaguamish, the Duwamish, the Sacramento, the Umpqua, or their tributary streams—I sense the river as a presence that takes form from specific, yet constantly shifting physical, social, and psychic forces. I first began training in sensing and interpreting these land-water histories while staking rocks and digging swales on a stream restoration crew.[15] I learned how to see broader geomorphic patterns and geologic processes in deep time along the glacial rivers of western Montana.[16] Later, training in ecology let me see the ecological relations among the organisms that live in streams and the adjacent riparian zone.[17] In one sense, the river is water's rhythms and flows in flood and drought, both the surface flows and the flows underneath.

In another sense, the river is the infrastructures and policies that amplify, diminish, or divert river flows. Considering a river this way reveals it as a sociotechnical system of interacting political and social forces that shape those flows.[18] Such forces include property regimes, regulations, management policies, incentives, rules, and tacit norms of use. If we consider water governance as a conceptual river, then the surface flows are the settler sciences, policies, infrastructures, and institutions that govern and direct river and water management. Just as one might hone an ability to sense fluvial and ecological histories by learning to see abandoned channels and old river terraces while walking across a floodplain, one can learn to sense political and social underflows by studying resistance to top-down management. Native nations and marginalized communities contest property doctrines and laws that legitimize groundwater overdraft and floodplain destruction and that interrogate scientific results with Indigenous and land-based knowledge. These underflows, these often-unnoticed resistances to extractive river management,[19] interact in complex ways with top-down water governance.

In conversation with interlocutors from land-based communities, I propose that other-than-human beings can also resist enclosure and damage caused by settler-colonial actions. Beavers and willow trees and elementals like landslides and floods all disrupt human river-control schemes. Rivers display the stochasticity and ungovernability that Native sciences have long recognized as the source of ecological diversity and abundance. Rivers inscribe the land with evidence of alternative histories and hint at latencies that their flows might awaken. Rivers also inspire their creatures with a sense of direction co-present with unruliness. Figuring this unruliness as resistance to Manifest Destiny projects opens up theoretical space for solidarity beyond the human.

The concept of underflows unites all of the disparate concerns of this book. But what are underflows? The word translates a key concept in hydrology, the hyporheic zone in and under a stream's bed (from the Greek *hypo*, under, and *rheos*, flow). The surface flow of a river is the part that is seen from the shore or a boat and that includes the undercurrents that roil its shimmering surface and the deep currents that can suck a swimmer under. Underflows are the parts of a river's flow that can't be seen: they seep through gravel or soil or rise from deep aquifers through cracks in bedrock to surface as springs. Underflows are this book's central concern, both in their watery materiality and in the discourses and politics that shape river governance. Just as the water in the hyporheic mixes with a stream's surface flow and deep groundwater, conceptual underflows percolate and influence conceptual surface flows of settler-colonial river management and the mainstream ecology, hydrology, social science, and economics that shape it. I break open the theoretical and associative potentials of "underflows" by exploring latent discourses that lie hidden or disregarded, like fluid underflows are hidden beneath a stream's bed.

When we are sensing in this physical and conceptual mode simultaneously, rivers come into view as messy riparian braids of life, water, and sediment, shaped by and shaping human societies as they practice interfacing with other animals and altering flows using science and other systematic practices. This book invites queer trans thought as a strategy for transfiguring settler-colonial riverine science into a tool for justice, and it makes a space for queer trans bodies and ways to live together.[20] One of my main projects is to think with water's and rivers' materiality about the political and social dynamics that affect who governs rivers and what sciences they use to do so. The underflows strategies I present here make a politics of solidarity with water and rivers, by understanding them as willful, unruly, feeling, acting beings. Underflows strategies ground that politics in alliances with Native nations' sciences and management practices. I shape this inquiry around three questions:

1. What politics of solidarity can queer trans feminist tactics bring into river science and management? This first question provokes ecological sciences not to be so straight by theorizing how scientists can mobilize queer and trans strategy to combat white supremacy and settler colonialism in environmental science and river governance.
2. How might queer and trans modes of relation among and beyond the human reshape science and governance along rivers? This second question invites queer and trans theorists to take seriously people's specific relations with the other-than-human world, and to consider scientific ways of knowing in riverine sciences as sources for queer and trans thought.
3. How might taking science and field practice seriously as a grounds for queer and trans ecological thought reshape queer, trans, and environmental studies? This third question challenges ecocritics working with queer and trans ecologies to center queer and trans lives and our relations with other-than-humans, and to ground queer and trans ecological thought in Black, Indigenous, Brown, and disabled queer and trans relations to the more-than-human world.

These questions theorize queer-trans-feminist river sciences as underflows that mix, mingle, and sometimes ally with Indigenous and environmental justice practices of science. Underflows, as I present the method, is not a unitary discipline or a strategy. Rather, it is a practice, an orientation, and an invitation to attend to hidden flows and their movements, excesses, and relations. The underflows method is of and for movements that protect and strengthen well-being in the face of settler state violence, which attempts to erase ways of living otherwise.

How River Scientists Sense Physical Underflows

I learned rivers by swimming and underflows by walking dry streambeds and sensing for paths of hidden flow. This flow makes relations among species, geographic features, and elementals like rain, snow, flood, drought. These relations are the proper object of normal science, but that science also includes embodied social relations among ecologists and their coconspirators. Though I swam many rivers before this and also walked the dry beds of many streams, the underflows approach first coalesced on one minor creek in the Central Coast of California, where I first swam, walked, and sensed in an ecological way.

On Salmon Creek, over three months of foggy-sunny summer, I walked alone or with other students every few weeks for seven years, into and through a time of deepening drought. We walked to observe and measure the stream as it shrank from a chattery flush and swirl of flows around logs and down cobbly riffles to still pools separated by long, dry swaths of gravel over the rainless summer. Sounds also make up this relation: the sounds of walking—crunch of wading boots on gravel, swish of nylon waders, chatter of young naturalists spotting a bug or lizard; the sounds of water flowing, or the almost-silence of flow's absence; and the everyday sounds of rural life—trucks passing on the road, distant chain saws, pumps, tractors, or mooing cows, birds cawing or tweeting or hooting. We would pause our walk at each pool. In early summer, the pools stretched clear and deep to the next trickling riffle; by late July, they had shrunk to mucky puddles, brown with tannins or black with cyanobacteria that grew in the absence of oxygen (so our sensors said). Feelings and sensations also materialize this relation—the pressure and coolness of air on skin, the squeeze and sweat of neoprene, cool water seeping in through holes in the knees. Repetition—of measurements, of movements, of walking or wading or snorkeling—marks this relation's temporality. Repetition layered memories; it made the novel familiar but also allowed me to sense and note changes: a trickling riffle fades to damp rocks that then dry with algae, leaving an ashy crust. Golden alder leaves drift down leisurely to rest on the stagnant water in a shrinking pool. After the first fall storm, a tree fallen across the stream caught a huge pile of brush and made a new pool and sandbar. By noting and theorizing from these affective and collective relations, in the field, with human and other-than-human collaborators, I probe these relations—underflows to the surface flows of ecology papers and conference presentations—to show how differences of race, gender, Indigeneity, disability, and positionality matter in the practice of river science.

Ecology, as I practice it, begins in and returns to the field. The field site may be a rural creek, a dry streambed, a large river, or an engineered waterway straitened and straightened between riprap. In the field, I observe, measure, and sample, writing notes and sketching into a water-resistant notebook and recording measured values on handheld computers, GPS receivers, or paper data sheets. At times, sampling water, sediment, and organisms requires elaborate apparatuses and collective labor. To tag fish with PIT tags requires a crew of at least five people—two to catch them in nets and later release them, one to weigh and measure, one to inject the tags under their belly skin, and one to record data, as well as various scales, rulers, buckets, aquarium bubblers, anesthesia, nets, cameras, scanners, and the like. To sample a stream's underflows for analysis in the lab, I and an assistant drill holes in PVC pipes, then drive these monitoring wells into the gravel with a post pounder and custom-built driver. To continuously measure temperature gradients below the streambed, I devised an apparatus made of thick wire with three dime-sized devices glued to it fifty centimeters apart, marked with labels made of scraps of soda cans, then inserted it into this well. This kind of crafting, small-scale manufacture, and improvisation is ubiquitous among field scientists, who occasionally publish papers on how to repurpose devices from other branches of science or industry to sampling tasks. It is not something that nonscientists often think about as part of the practice of science.

Collecting field data on fish, water, or habitat is only the beginning. These data must be extracted, distilled, and visualized using spatial, numerical, or statistical models. For example, the temperature gradient apparatus I made makes use of heat as a "natural tracer" to trace water flow beneath the streambed. Water at the surface of a stream warms up in the sun and cools during the night, but a meter or so below the surface of the streambed, temperature varies little from day to day or season to season. Fishes make use of this gradient, seeking out groundwater-fed pools as cool refuges during hot spells. The pools also stay warmer than shallows in cold winters. By comparing the amplitude of daily fluctuations at the streambed and fifty and one hundred centimeters below the surface, I could tell when the stream was welling down to recharge the groundwater and when the groundwater was welling up to feed the shrinking pools. Analyzing these data with heat transport equations, I could model how much water was welling up and seeping down.

Hydrologists triangulate with different tracers to visualize subsurface flows. I and helpers collected samples of as many different "source waters" as we could access to characterize the "signature" of the aquifer—allowing us to trace the flow of groundwater into the stream. Since the field sites were on private land, gaining access to these wells and springs required a community effort, over three years. I began this effort by interviewing many people about their water source and water use. I met residents at a community potluck, and my collaborators at the watershed council introduced me to more. I also collected samples from monitoring wells in the stream every few weeks to understand how different waters ebbed and flowed over the course of the summer drydown. To collect these samples, my field assistants and I lugged a peristaltic pump down to the creeks and pumped water from the pools and monitoring wells into syringes, which we dispensed through plastic filters into small amber vials. These backbreaking endurance performances, staged over weeks and years on end, aim to answer questions that can help salmon live or die in these little streams where people also live. In this case, we wanted to know which reservoirs of groundwater fed into pools where water and dissolved oxygen persisted through the summer, enabling salmon to survive, and which residents pumped from these same aquifers for their drinking and household water. Knowing this, residents could try to conserve water to reduce their pumping or build alternate sources like rain tanks, strengthening individual resilience and community capacity for self-management. Conservation district staff could design ponds to store and recharge groundwater and use our data to apply for funding to build them.

Back at the lab, we zapped the vials with UV light, which fluoresced in indicative patterns to describe the molecular forms of carbon the clear water contained. Dissolved carbon from soil or algae hinted at the paths rain took over and through the ground to reach a pool. Some pools had lots of labile carbon, from leaves or algae, which fed blooms of bacteria that later died and decomposed. These bacteria, and the others that decomposed them after death, all consume oxygen, sometimes so much that fishes suffocated there. Other pools, fed by groundwater, lost oxygen more slowly, even after the quick-flowing riffles between them (which mix oxygen back into the water) had gone dry.

At Salmon Creek, during a four-year drought, these underflows diminished. As less rain seeped down through the soil into cracks in the metamorphic rock beneath, local residents tasted the change. Groundwater grew tangy with magnesium and iron that leached from metamorphic rock. Then some wells and springs went dry for the first time. With less groundwater to feed it, the stream stopped flowing. Water in the isolated pools that remained became stagnant, black, and stinky. Probing these underflows' chemical makeup and flow paths, I came to think of them as the river's memories and messengers. As flows moved from ground to surface water and back again, mixing all the way, they carried traces of fish, plant, and microbial life, of the sun and rain, and effluents from dairy barns, roadways, and septic tanks. If we scientists chose the right sensors to measure them and the right models to tease out patterns in the data, underflows' stories could help people make a future for coho salmon and steelhead trout amid rural sprawl and climate change.

Collaborative science, as I practice it, doesn't end with presenting the results of the above scientific investigations at an academic conference or with publishing a journal article. It returns to the field as a relation. Like Indigenous research methodologies, which require of the researcher reciprocity and ongoing relationality, queer ecologies make kinship networks through the practice of field science. I joined one such network in Salmon Creek, where I encouraged agency scientists and local residents to think about reconfiguring their everyday water practices into practices of care for rivers and salmon. Science is one such practice, but it is not the only one. Local residents' measurements and stories of their wells and springs, their enthusiasm for or rejection of my methods and conclusions, and the friendships we developed all became part of a collective practice of care for Salmon Creek. Years later, we still meet up when I'm in the area and catch up on conference calls about water policy. This approach has guided my way of working in new research networks in the Klamath Basin and Lower Duwamish River. Just as traces of carbon in water flowing underground can carry messages about how infiltrating rain affects streams in a biophysical watershed, theoretical underflows from Indigenous, feminist, and queer-trans STS can carry messages and visions for justice in riverine science. These messages show that place-based alliances are transfiguring how people manage water and relate to rivers. In the best case, these alliances actively work to return Indigenous land and strengthen sovereignty.

What Is Queer Trans Feminist Ecology?

As I live and think it, queer trans feminist ecology is an unfolding praxis, mood, and orientation to solidarity and multiplicity that holds space for different kinds of thought:[21] in the field and with other-than-human beings. This book traces and deploys underflows as a mode of thinking and strategizing in feminist science studies, a field that uses feminist politics and theories to critique and transform scientific practice, goals, and ethics. This book's object is queer and trans ecology, in two senses. First, underflows as approach *shows* queer and trans ecologies. Tracing conceptual and material underflows is a queer strategy that does the work of showing queer affinities—*queer* in the sense of resisting/refusing norms of cisheteronormative settler-colonial water politics and land relations. I figure these hegemonic politics and relations as surface flows, as in the part of the river that you can see and swim in, in contrast to the underflows that seep beneath the streambed unseen. Tracing underflows is also a trans strategy that displays trans affinities for transfiguring our human bodies as we make collective structures of support and celebrate one another's bodily and affective transformation. As with queer affinities, trans affinities can produce and inspire new politics of solidarity with rivers and their relations: human river workers and other-than-human beings join together in the subsurface and latencies of Manifest Destiny projects to overturn their methods and logics.

Second, underflows *are* queer trans ecologies, made by queer and trans people. Our ways of being together along shorelines and in the water involve, often and crucially, river and waterfront spaces. At times, queers—especially queer trans scientists—can find connection to nonhuman ecologies and the world when cishet-settler worlds reject us and our deviant relations. Queer kinship can remake damaged ecological relations. Trans future-thinking and our experiences of transformation can help grapple with an unpredictable climate future-becoming-present.

What does queer-trans-feminist approach *do* for and to science and for river management, policy, and politics? In other words, how, precisely, do queer and trans people's lives, thoughts, and ways of being relate to and matter for rivers? Can queer trans approaches to ecological science help transfigure settler relations to rivers and watersheds? As I elaborate in this book, queer trans feminist ecology is a way of asking questions with others, grounded in everyday lives and livelihoods, and attentive to ongoing projects of settler colonialism and white supremacy. This way of doing science rests on a politics of solidarity, in which scientists follow protocol and respect knowledge practices developed by frontline communities, while cogenerating research that meets community need and builds community capacity.[22] Queer trans feminist ecology is committed to radical collective action to transform settler institutions and relations to rivers, to center Native sovereignty, and to bend environmental projects toward justice.

Tracing resonances among Indigenous, queer, and trans theories of kinship, reciprocity, consent, and collective action, I follow a path cleared by scholars, fishers, and river workers who argue that Indigenous sciences are essential to recovering health in riverine lifeworlds.[23] Like Kim Tall-Bear (Dakota) and Zoe Todd (Métis) assert for genomic sciences and fish management, respectively, I hold that queer trans feminist strategies for ecological well-being must center Indigenous theories of relationality. These theories are finely honed for transfiguring settler-colonial practices of science into a mode that protects and recovers rivers and landscapes so that they provide capacities that support the collective continuance of Native peoples. As defined by Kyle Powys Whyte (Potawatomi), a collective capacity, such as "salmon habitat ... entwined with human institutions," supports Indigenous self-determination, yet "settlement works to erase this capacity through diverse actions ranging from treaty violations to the ignorance of private citizens whose actions, such as littering or pollution through their business ventures, add up to degrade salmon habitat."[24] Whyte argues that settler collective continuance always takes from Indigenous collective continuance. Thus, for Native nations to recover capacities that enable their collective continuance, settler-colonial river relations—from infrastructures to floodplain management to governance and policy to settlers' collective values— must change.

To effect this change, river scientists and managers, particularly settler scientists who dominate universities and agencies, must listen to path-making conversations on grounded solidarity and work in respectful alliance to join and apply them to natural resources management and restoration.[25] Conversations in Native American and Indigenous studies have provoked crucial insights in Black, Postcolonial, American, and Canadian studies, yet most environmental scientists and managers don't know this work. River restoration integrates Western scientific disciplines and sometimes incorporates Indigenous and local knowledge into scientific analysis,[26] but I propose a deeper transformation, rooted in Indigenous and feminist science studies analysis: decentering Western science as an objective arbiter of truth. Western science does not have a monopoly on structured observation, long-term study and record keeping, or repeatable analysis of data. Indigenous, land-based, and urban people also practice systematic observation and analysis, sometimes reappropriating Western science techniques.[27] The feminist science studies concept of "partial perspectives" highlights that each way of knowing is embedded in assumptions and values, giving it power to answer some questions but obscuring others from even being asked. In river restoration, mutual listening and comparing results from tribes', agencies', farmers', and fishers' partial perspectives can sometimes produce a stronger objectivity—stronger because it contains a more robust understanding of historical variability in floods and droughts or because it includes a wider range of organisms, processes, and human activities that trigger changes in ecological states.[28] At other times, partial perspectives conflict, as when subjugated knowledges challenge Western science data on human and ecological health.[29]

For ecological sciences practiced in settler states, Indigenous approaches provide a critical partial perspective, a perspective that is often devalued by non-Native scientists and institutions. Indigenous research methodologies are at the forefront of modeling research ethics, protocols, and alliance-building across different forms of knowledge.[30] Tribal communities often find their traditional practices and knowledges dismissed in state and federal dam removal and climate adaptation governance processes.[31] This exclusion of underrepresented voices from shaping scientific agendas and translating findings to policy impairs progress by (1) minimizing epistemic diversity, thwarting more rigorous questioning and stronger objectivity; (2) perpetuating injustice in frontline communities and limiting policy reach by neglecting research questions related to health, livelihood, and ecological well-being; and (3) rejecting "unconventional" methods for finding solutions and connecting knowledge to action.[32] By privileging dominant ways of thinking and working together at the expense of Indigenous, feminist, and other marginalized knowledges, the "best available" science will continue to reproduce a narrow, Eurocentric cultural construct. Science as multiple disrupts settler science's presumed universality and objectivity, building solidarity with Indigenous sciences and land-based knowledges that are explicitly grounded in alternate ontologies and ethics.

Though my empirical research on salmon and river management is deeply embedded in laws, policies, and institutions of Native and settler governance, this book does not systematically survey water governance, treaty rights, history, or ongoing legal battles. Through the cases in this book, I demonstrate how thinking sciences as multiple complicates and challenges Manifest Destiny logics.[33] Studies that begin from and center land-based knowledges and Indigenous protocol create different research questions, methods, and ethical frameworks than do studies designed from afar by settler academics or agency personnel, which discount Indigenous knowledge of ecological trajectories and cultural protocols for restoration. I present examples from my own research on beaver and salmon habitat restoration, through collaborations with Native nations and grassroots groups, to show how research grounded in multiple sciences and queer-trans strategy changes river management.[34]

I also explore queer and trans people's kinship and solidarity networks as a model for caring for other species and damaged places. We care for one another when rejected by our families of origin. We mourn together when, disproportionately, trans women and gender-nonconforming Black people, Indigenous people, and people of color die at the hands of police and bigots, or because they lack health care, housing, and social services. Sometimes, when queer or trans care networks can't stave off pressure from mainstream society, we lose chosen family to despair and suicide. At vigils and memorials that sometimes turn into street protests, we celebrate one another's flamboyant brilliance in creating life and love against obliteration, splashing in the waves wearing the sexiest swimsuits. These practices produce relations and energies that Elijah Adiv Edelman has theorized as trans vitalities: "radical care, or the care offered outside, beside, underneath, and perhaps even above normative outlets, is . . . care that follows desire lines. [Capitalism's] discourses prevent, rather than facilitate, a grounded celebration of desire lines, of explorations of livable life, of the unimaginable possibilities in approaching all life as sacred. What I am proposing is, instead, a profound and radical disinvestment of hierarchies of worth."[35] Edelman, here, echoes Taylor's rallying cry, "We want to live!" as an expression of trans collective desire for Black trans women's vitality. In highlighting trans vitalities' power to dismantle worth hierarchies among humans, Edelman hones in on desire as a trans tactic that, growing from our lives and collective praxis, celebrates the "unimaginable possibilities in approaching all life as sacred." If late capitalism constrains the imagination of such lifeworlds, queer/trans-of-color and Indigenous scholars *do* imagine many possibilities, grounded in Native and activist desires and ways of being together.[36] Edelman's exploration of "all life as sacred" can extend beyond the human to include other-than-human relations that are devalued in mainstream society.

This confluence of vitalities, care, and resistance *embodies* and *makes* queer ecology/ies in two senses. On the one hand, I explore queer ecologies as actual geographical places and probe how to unearth and understand queer and trans people's use, inhabitation, or passage through them. On the other hand, I approach queer and trans ecologies as a method. This method deploys queer tactics for projects that situate and critique scientific thought with art and political thought in order to imagine science in a new way.

Site: Underflows in Streams and Politics

Rivers flow through cities and rural areas, carrying traces of livelihoods and hosting migrations of species. As human-added toxins seep through underflows, they, like the "natural tracers" of dissolved organic carbon and radon gas I used in Salmon Creek, can be storytellers and messengers. In the San Francisco Bay, underflows seep from gold rush–era mercury mines in rural pasture lands outside San Jose into tributaries that dive into culverts under Silicon Valley, and they feed into Guadalupe Creek. Where the creek flows into San Francisco Bay, mercury concentrates in the mudflats where methylating bacteria transform it into a damaging bioavailable form. Unhoused people make their homes on the banks and are exposed to mercury, sewage, and a stew of other toxins that wash off of city yards and streets. Mercury concentrates in living tissues, accumulating up the food chain as mussels and clams filter plankton and algae and as small fishes eat plankton and are eaten in turn by larger fishes. Now mobile, mercury moves to fishing piers around the urban shorelines of San Francisco, Oakland, and Richmond and into the bodies of subsistence fishers and their children (predominantly Black, Latinx, and Southeast Asian people), causing bodily harm. Some fishes swim out to sea, where tuna and marine mammals eat them. Toxins accumulate through the food web. Similar webs of relation bind together old industrial sites and Indigenous and subsistence fishers on Seattle's Duwamish River, where I now work, as PCBs and heavy metals from aviation and manufacturing linger and harm human and other-than-human bodies. Thinking these relations as underflows invites scientists who work to remediate toxins into responsibility and response-ability—to support community demands for justice and to confront settler-colonial and racist policies that perpetuate these harms.

This story of connection across artificial boundaries of urban and rural characterizes all of the western US rivers that are this book's conceptual and physical sites. A river system does not only comprise the flows and movements of water across the landscape but is also produced by specific histories of extraction that have altered these flows in hydraulic societies.[37] Rivers inspire politics and conflicts focused on water's paths down channels, through the subsurface, and through diversions in pumps and canals. Fishes, birds, and invertebrates swim in rivers that people use for navigation, irrigation, pleasure, and fishing. Rivers that overheat or run dry, sewers and agricultural drains that discharge into spawning grounds, and fisheries closed year after year as populations plummet—these rivers display severed relations between people, fishes, and the waters they swim in. Such are the legacies of nineteenth- and twentieth-century settler projects to dam, divert, and confine free-flowing rivers. Rhetorics of improvement and control of unruly nature persist in settler law, policies, and institutional structures.[38]

Settler policy confines these rivers conceptually as well, into classifications of "wild and scenic" or as a resource for hydropower, irrigation, municipal water, and waste dilution. Writing of the Columbia River, historian Richard White uses "work" to describe both the river's energy to move water and sediment across its own floodplain and the Native and non-Native labor of building dams and levees—to transform the river into a hybrid "organic machine."[39] But because settler control is never perfect, the river became a mixture. Organic elements, including workers, fishes, and the water's pulse and flow, resist and exceed machinic rationality. In recent decades, river engineering—this mechanic rationality—has incorporated ecological and social science approaches. This multidisciplinary riverine science is one object of this book's analysis. In US settler-colonial contexts, rivers' futures depend, partly, on science, because science is bound up with their law, regulation, engineering, and management. This work continues as land-based communities mobilize for dam removal, Superfund cleanup, and river restoration. If "what's real is the mixture," an underflows approach adds into the mix latencies that stir beneath the abstraction of water.[40] Two senses of abstraction resonate here: diversion from a source and Jamie Linton's concept of water as abstract, purified into infinitely exchangeable H_2O that can be pumped from place to place without concern for the lively multispecies worlds it once sustained.[41]

In sum, rivers are not only rural or urban but transit through alternating gradients of urban and rural and suburb. The river is not only ground or surface water but the seepage back and forth, not only material or political but also shaped by feelings and memories of people who visit and work them. Muddying the waters of Manifest Destiny projects and the science-as-usual that produces them, *Underflows* queers political ecology analyses of the hydrosocial—this separation of flowing and working rivers that river sciences inscribes.[42] For example, applying Muñoz's brown commons dissolves human/nature binaries on industrialized rivers, as I will show. An underflows approach understands these sites to always be embodied, political, personal, and relational.

Queer Trans Science as Underflow

I define queer trans approaches to river sciences as one underflow to mainstream science's social and affective practices. When cishet-settler worlds reject us and our deviant relations, queers and trans people often find solace in connection to nonhuman ecologies and the world.

For queer and trans scientists who are marginalized and isolated within the majority white, straight, cisgender spaces of river science, grounding scientific practice in our theories, knowledges, and relationality is a matter of survival. I explore in these pages how queer and trans ways of being together—which often unfold in river and waterfront spaces—generate powerful scientific approaches for understanding queer ecologies that include human and other-than-human beings. These approaches, joining in solidarity with underflows of other marginalized sciences, can well up and change river science. More broadly, queer and trans tactics—honed by amplifying queer and trans people's voices within environmental and climate justice movements—can deepen solidarity across campaigns for dam removal, toxic cleanup, and ecological resurgence.

Bridging queer-trans and ecological theory is no simple task. First, queers, including humanist scholars of queer ecologies, and ecologists talk past each other. Humanists have neglected environmental science and ecological theory as a grounds for their work, while queer and trans scientists rarely read queer and trans theory in literary or performance studies. Second, because we queer and trans ecologists often compartmentalize these identities, we often don't imagine that our cultural knowledge and embodied experience have anything to do with science, politics, and management.[43] Third, queer and trans theorists rarely intervene in environmental studies or management. Instead, queer and trans theory and cultural production that explores ecological relations often invokes ecology as a metaphor, rather than a scientific discipline. But we queers and trans people who become scientists know uncanny kinship, transformations, change, and how to live fabulously while facing systemic attempts to destroy our life chances. Some of us also bring political analysis and organizing strategies learned while organizing for sex workers and immigrant rights, racial justice, and prison abolition. By bringing this knowledge inside of ecological theory and field practice, we can model for straight scientists new ways to resist ecocultural destruction.

Is queer trans ecology—in the sense of ecological science done by queers or trans people—distinct in its working methods and goals from straight ecology? This question is crucial both for queer and trans scientists, who face microaggressions and barriers to full participation in natural sciences, and for straight and cisgender allies, who hope to erode those barriers. Black, Indigenous, and immigrant ecologists have argued for more inclusive science to increase relevance and excellence. Native and environmental justice groups have focused environmental management on food, health, and access to land for cultural continuance. Queer scientists, farmers, and environmental practitioners have always been a part of environmental justice movements, yet they rarely characterize their work in ecological science or with ecosystem repair as queer or trans acts.[44] Unlike Black ecology and field biology, which has a growing literature on field methods, analysis, ethics, and pedagogy,[45] queer and trans ecology has no distinct literature on methods yet.

While rivers contain queer and trans people who swim, fish, relax, and work in and along them, queer and trans people's ecological relations are hardly ever considered in river policy. Queer and trans histories and literatures of belonging along rivers and shorelines most often take place within cities, while most straight settler histories figure fishing, logging, ranching, and farming as cisheteronormative.[46] Popular culture shows queers and trans people in urban settings as producers of culture, as if the unnaturalness of our desires and bodies places us outside a pristine Nature. Yet even within urban spaces, parks and restoration projects are often designed to exclude queer and trans activities like cruising and sex work, by eliminating brush and installing bright lights. A queer aesthetic intervention against this "family friendly" (read: cisheteronormative) planning could design parks and waterfront spaces to create habitat for riparian species that also provides the secluded nooks and sightlines that make for a good cruising park.[47]

Like recent rural queer trans production, I want to resist a caricature of us as (only) urban, by considering Native and non-Native queer relations in rural landscapes.[48] Queer land projects—some reconfigurations of Radical Faerie sanctuaries and Women's Land projects of the '70s and '80s—have become spaces of contact for queers and trans and nonbinary people to practice forestry, farming, wildcrafting, botanical medicine, butchering, and stream restoration.[49] These spaces often erase or appropriate Indigenous relations, reinforcing white supremacy and settler-colonial logics in their cultures and structures.[50] Rural queer ecocultural practices are a rich site for theorizing what queer ecology could mean for regenerating degraded rural ecologies and livelihoods, while confronting anti-Black racism and working with Native nations.[51] Here, however, I theorize queer and trans land relations from a different angle, not (primarily) through living in rural areas, but through ongoing practice of doing field science, mostly in rural communities.

In sum, if rivers are both machine and mixture, shaped by settler fantasies of control embedded in river sciences, then queering and transgression can unlock other potentials of ecology and suggest new lines of flight toward unsettling riverine relations. In the chapters that follow, I elaborate this argument by exploring the effects of queer trans feminist interventions in river science and management practice, proposing ways to nurture queer and trans scholars and river workers, and grounding queer ecological thought in Black, Indigenous, Brown, and disabled queer and trans lives.

Definitions

Within this project of bridging queer, trans, feminist, and ecological theory, *restoration* and *recovery* require close scrutiny. I use *restoration* to refer both to the resurgence of salmon populations and to river engineering practices. That Native and non-Native managers still favor the term *restoration*, despite critique from academics, suggests that its many definitions can create pragmatic convergence among theoretically divergent projects. *Restoration* is often used as an umbrella for other "re-words"—*rehabilitation, rewilding, reintroduction, reconciliation, repair*, and so on—that have been critiqued from all sides yet still persist in scientific and practitioner circles.[52] Some critics challenge the settler-colonial doctrines of *terra nullius* implicit in restoration to a wilderness state.[53] Others note that given the scope of anthropogenic changes to lands and waters, terms such as *reconciliation, rehabilitation*, or *intervention* are more accurate, particularly where environmental justice and restoration intersect.[54] I sometimes use *recovery* following Whyte and Billy Frank Jr., who argue for salmon recovery as a return to health that includes but also is more than tribal members' individual recovery from health problems caused by lack of traditional foods.[55] At other times, I use the term *rehabilitation* to indicate ongoing adaptation to damage that can't be reversed—what others have theorized as reconciliation ecology or ecological reparations.[56]

Defining "science" as multiple forms of systematic knowledge is also key. STS scholars have long argued for provincializing mainstream science, using adjectives like settler, Western, normal, or Royal or distinguishing Euro-Western Science with a capital S.[57] In river sciences, and environmental sciences, settler and Native resource managers frequently share data and collaborate on species conservation, harvest, and ecosystem restoration through comanagement processes rooted in treaty rights. Yet official settler accounts of environmental sciences almost always erase Native origins and ongoing innovation and leadership.[58] Building on their early exchanges with settler scientists, tribes and First Nations have continuously innovated scientific practices like measuring, mapping, and statistical techniques within an Indigenous epistemology. By naming Indigenous knowledge practice *science*, I join other Indigenous and feminist scholars in rejecting the settler binary logic that says that Indigenous peoples cannot use "modern" methods while maintaining "traditional" practices. In this book, *Western* or *Indigenous* descriptors of science clarify different actors' contributions to coproduction of science and regulatory regimes. Unmarked *sciences* sometimes indicates a systematic approach practiced in Western and non-Western contexts.[59] I never use unmarked *science* or *sciences* to mean only Western science but specify Indigenous, settler, farmer, or lay science. I do this to emphasize that river governance often mashes up data and models that are rooted in settler agency, community, and Native natural resource science.

Indigenous science as ongoing and resurgent practice, queer and non-white modes of kin making, and collective governance in settler communities: potent in combination, these underflows demand that riverine science become more relational and responsive to marginalized peoples' experience and knowledge. Conceptual underflows of Native and grassroots vision and strategy for river renewal are often invisible to—or disregarded by—federal and state managers. Conceptual underflows circulate at protests against pipelines and pollution, fish-ins challenging anti-Native regulations, court battles, and everyday practices of caring for and harvesting from rivers and floodplains. Tracing underflows reveals cracks in tacit settler-colonial doctrines of river control. Making underflows the grounds for scientific, political, and social action on behalf of riverine worlds strengthens possibilities for Native–non-Native alliance around river renewal. When these alliances revive suppressed possibilities for relating otherwise to rivers, underflows come into view.

Methodology: Following Water's Phases

Inside clouds, raindrops condense around nuclei—a speck of dust, a crystal of salt, a mote of ash from a forest fire or power plant. High in roiling cumulus thunderheads or low in dark nimbus, temperatures drop and vapor molecules turn to liquid on the surface of this small particle. These droplets collide, growing larger and heavier until they mass enough to drop toward the ground. A young Samuel R. Delany flies through such clouds at the close of *The Motion of Light in Water*, on a propeller plane over the Arctic, writing in his journal, "I looked out the window at walls of moonlit cloud rising beside us as though we were at the bottom of some gray and ivory canyon, hung above the moon-smashed sea." An older Delany speculates that this watery image captures a disorientation that characterizes his young queer Black writerly life on the cusp of Stonewall:

> With whatever hindsight, I suppose the reason that I want to close on a consideration of these words is that the moon-solid progress through high, drifting cumulus is—read them again—at the very opposite of what we perceive on a liquid's tilting and untilting top, and so becomes the other privileged pole among the images of this study, this essay, this memoir.
>
> Or perhaps, as it is only a clause whose syntactic place has been questioned by my own unscholarly researches, I merely want to fix it before it vanishes like water, like light, like the play between them we only suggest, but never master, with the word motion.[60]

I reread *Motion* during pandemic isolation, as rage against anti-Black police violence condensed into collective uprising, and flashed on the physics of condensation as another pole of indeterminacy and state change to what happens inside the clouds Delany flew through. This image of condensation, infused with a feeling of disorientation that Delany's writing on queer memory generates, inspired a structure that describes this book's methodology: a nucleus of collaborative praxis rooted in Indigenous and community protocol and story, surrounded by condensate of archival, ethnographic, and ecopoetics methods. Like the images that make up *Motion*, which present emotional or perspectival registers of a moment by overlaying different time lines, these methods influence one another, layering like thin films of water in a raindrop as thought and theory arise and morph across sites and field visits.

[Raindrop Nucleus] Indigenous and Cogenerated Research Methods: Centering Community Needs and Strengths and Figuring Out Situationally Appropriate Methods

In my work on rivers, I cogenerate research questions and approaches with tribes and community-based organizations. Together, in a process that unfolds over months or years, we discuss which Western ecology and hydrology techniques we will retool. and what community-grounded protocols will guide our process. We then come together to discuss results, critique models, and plan further action. My approach draws from Indigenous research methods, participatory action research, and critical development studies. I undertake these collaborative projects with a queer ethic of solidarity and a trans ethic of transfiguring normal ways of doing science. This approach centers community health, planning, and culturally relevant economic development, pushing river science and restoration to deeply engage power and collective continuance.[61] Indigenous research methods unfold through Nation-specific protocols that tie people's well-being to that of the land and water. When I and others cogenerate research in non-Native communities, we often refuse to separate the ecological from the social. In grassroots environmental justice campaigns, science's authority is not universal or unquestioned, but neither is science dismissed out of hand. Rather, we make science serve the people and community well-being, not only through its research outputs but by creating pathways and funding for Black, Brown, and Indigenous people to work in science and watershed repair.[62]

As I support this frontline work, ethics of reciprocity, community benefit, and respect for multiple ways of knowing shape the study designs that I codevelop with collaborators. When I do river mapping, water quality and flow studies, and salmon surveys, I also analyze the social and power dynamics of top-down collaborations, using a queer trans feminist science studies praxis grounded in field interviews and community-based workshops. Over seven years of work on Salmon Creek, this praxis produced new ecological knowledge, an ongoing community-led wet-dry mapping project, new bottom-up methods for studying household water use adaptations during drought, and collective reflection on water's value for agriculture and salmon-based livelihoods.[63] Feminist STS praxis that valued local knowledge and grounded research questions in different residents' and regulators' priorities built community support for my groundwater and stream studies. Without active collaboration from local residents, such studies on private lands would have been impossible to execute. Had I not framed local knowledge as crucial expertise for drought and climate adaptation, reflection among residents and agency scientists in collaborative workshops might not have produced lasting partnerships—in well monitoring, wet-dry mapping, and groundwater recharge. These approaches and studies focused on a small (ninety-square-kilometer) watershed but have informed federal salmon recovery and state water planning efforts regionally.

In my newer work in the Mid-Klamath river, I work separately with the Quartz Valley Indian Community and Scott River Watershed Council, and with an existing collaboration between the Karuk Department of Natural Resources and the Mid-Klamath Watershed Council. Each of these groups works independently with federal and state scientists and regulators, but they don't all work together. To increase contact and traffic in ideas among these groups, I cocreated an ecocultural restoration field school with these partners, co-led by Lisa Morehead-Hillman (Karuk) and the Karuk Department of Natural Resources. Interactions during the field school have brought people together across political divides, in informal settings like stream surveys and barbecues.

As part of a larger shift in river science and management, Indigenous and environmental justice research methods demand reciprocity and ongoing relationality from outside researchers. As part of my commitment to reciprocity and ongoing relationality, I do work that sustains my collaborators' day-to-day campaigns—making maps, helping write outreach and curriculum materials, co-organizing community forums. I meet with community members individually to discuss questions and concerns about climate, drought, fish, and water policy. These everyday practices are the nucleus on which my relational field practice condenses.

[Condensate] Ethnography of Water and River Politics: Kitchen Table Strategy Sessions, Field Interviews, and Coproducing Ecological Field Studies

The kitchen table, a hallowed site of feminist strategizing, is where my field ethnography often begins and returns to. Before heading out to count fish or map streams, we drink tea and coffee, eat manzanita pancakes or homemade cheese, gossip, and discuss logistics. The idea that science can be feminist, queer, or trans was new to my Salmon Creek and Scott Valley collaborators, but they came to appreciate the collaborative and relational methods I mobilized, and they grew curious about queer kinship and trans affects as ways to understand the social dimensions of their habitat restoration efforts. To come to these kitchen tables as a queer transmasculine person sometimes feels like a terrifying risk. Yet, at kitchen tables in Scott Valley and Salmon Creek, watershed council members and I have had frank and personal conversations about homophobia, racism, and anti-Indigenous sentiment in their communities. Scott River Watershed Council leaders (both women) discussed gendered dynamics of restoration politics in the valley, where implicit sexism and explicit homophobia and transphobia are common. Later, they assured me they would defend me and my students against homophobic, transphobic, or anti-Native attacks. Queer trans visibility in such spaces, made safer when cis and straight locals act as allies to queer and trans researchers, can create a little more breathing room for scientists of nonnormative genders and sexualities. This breathing room is one outcome of queer-trans-feminist ethnographic methods.

Kitchen table ethnography as I practice it is a counterpoint to field ethnography, which includes interviewing people about their water sources and uses or writing up observations of doing fish counts and water monitoring together. These complementary methods produce data on people's conceptual models of how water use affects ecology and human livelihoods, their opinions on river policy and politics, their ideas on how to study and fix degraded riverscapes, and their feelings for other species and the land and water. Whereas field ethnography encompasses single interviews and sustained episodic engagement, kitchen table ethnography can queer collaborative science by simultaneously building relationality and strategizing future actions.

[Condensate] Analysis of Affect and Subsurface Politics: Queer Trans Critical Physical Geography

To analyze the collective impact of these individual perceptions and feelings on river politics, I employ queer and trans analytical approaches within the framework of critical physical geography (CPG), which focuses on how power, labor, and political economy shape environmental science and management. This analysis reads ethnographic data and archival sources for people's perceptions of how settler-colonial structures shape water and ecology institutions and their experiences in collaborations across racial, Native/non-Native, and class lines. Such partnerships and research ethics are a critical concern for the field of critical physical geography. To demonstrate how Indigenous Studies theory and method can strengthen CPG's analytical power, I incorporate analysis of tribal sovereignty and settler colonialism to understand conflicts over water and beaver-assisted restoration. Crucial to this intervention is Michelle Daigle and Margaret Marietta Ramirez's contention that the decolonial in geography is "an affirmative refusal of white supremacy, anti-Blackness, and the settler colonial state," which "must form 'place-based constellations in theory and practice.'"[64]

I also take up Muñoz's cruising methodologies to describe a trans strategy for unsettling watershed restorations through collaborations with beavers that disrupt power dynamics and political economy of the restoration sector.[65] Deploying another Muñozian approach—ethnography of the trace—I developed a queer tactic that scientists can appropriate to mourn beloved species facing extinction. Where Muñoz read performance ephemera, New York streetscapes, and gestures in punk and gay nightclubs, I read scientists' gestures at ecology conferences, my own trans field encounters, graphs in my ecology papers, and performance collaborations through the art-science collective the Water Underground. Informed by feminist standpoint theory, these Indigenous and queer-trans approaches to document analysis and public scholarship sharpen CPG's analysis and produce stronger objectivity in riverine sciences.

[Condensate] Ecopoetics: Spatial-Narrative-Image Interrogation

The ecopoetics approaches that I and July Hazard have developed in field courses and public workshops overturn assumptions of natural and social science analyses, in the way that an underflow might undermine a street, creating a sinkhole that then makes people detour. Ecopoetics methods open space for new ways of observing, recording, and responding to the living world. Central to this ecopoetics is close observation and attunement through field writing, which is a core practice in the arts, in field sciences, and in STS. Hazard and I convened a countercanon of Black, Indigenous, immigrant, and diasporic poets, reading their work for relations that undermine colonial and settler-colonial ecological thought.

In field writing activities during these classes, Hazard and I ground the field as a queer space and develop writing prompts with a trans flair for making the familiar process of recording and enumerating data strange. We do this by cultivating new perceptual tools, embracing disorientation, and attuning to unpredictability and dynamism, in the guise of stochasticity. This underflow of method, elaborated later, includes the practice of making or finding an apparatus to sense buried flows, appropriating the ecological field transect as a method for plumbing memory through place-relation, and using diffraction patterns generated by tossing a rock into the water to invite questions and tactics for reimagining relations along shorelines.[66] When we and our students undertake these investigations collectively, we produce individual work but also collaborative poems constructed from one line of each person's poem, and we erect temporary waterfront galleries that invite passersby to reconsider their own shoreline relations.

From Sky to Soil, Interflows to Underflows: How Water Moves

To see how methodological underflows move through networks of river restoration and reimagination, I turn to water itself—to its movement, unseen, under the surface of the ground. It picks up traces; it dissolves carbon and minerals. As rain, it picks up acids from the sky and, carrying them into cracks in granite or voids in soil, dissolves the molecular bonds holding silica together with potassium or sodium as feldspar, weathering rock to clay. It moves down until it reaches the water table, an aquifer, an underground lake held in tension between grains of sand and gravel, or elsewhere in karst caves. There, the underflows from different tributaries or subsurface streams mix, react, recombine.

My underflows approach is grounded in the commitment that environmental science findings should contribute to queer and trans life, in part because queer and trans environmental scientists produce some of those findings. This approach generates new kinds of questions: Does a field study attend to queer relations, contribute to queer politics, or make it possible for queer and trans scientists to work where they couldn't before? Can we queer x trans x feminist scientists escape erasure and toxic masculinity/femininity and instead work in our brilliance and collective strengths? Feminist STS, as a core practice, attends to differences gender makes, on the one hand to the scientific methods, results, and actions taken from those results, and on the other hand to how gendered sociality influences relations in scientific spaces. This attention has gestated more egalitarian lab practices and new ethical findings and scientific foci.[67] Queer and trans methods deepen feminist STS provocations to science (normatively straight, white, and cismale) by asking what queer or trans relations mainstream scientific results exclude. Queer and trans people's embodied experiences of deviance, transition, scrappy livelihoods, and antiauthoritarian political acts inspire new scientific practices and environmental politics. Using underflows methods to channel those experiences, I have produced art-science works that circulate among river scientists and sometimes lead nontrans, nonqueer scientists to remix straight science with strange kin. Where water in a river exerts pressure on the aquifer, that pressure can cause an upwelling further downstream. To work in an underflows way is to see, recognize, and reach out for different streams—dissident streams, decolonizing upwellings that rework science and governance from within but also from below—into a lively, muddy, organic machine.[68]

Chapter Overviews

Situated in these rivers and movements, I model for feminist STS scholars how thinking in the overlays of Indigenous, queer-of-color, and trans thought can expand our field's objects of study and enrich our theoretical conversations. For river scientists and river workers, I model how confronting racism and settler colonialism with queer-trans tactics can begin to dismantle exclusionary structures that have long plagued Western field sciences and river governance. The first four chapters alternate between eco-hydrological underflows along rural rivers and conceptual underflows to science and field practice; the fifth chapter integrates these senses of underflows along an urban waterfront. In the conclusion, I present a research prospectus for my new work with the Karuk Tribe, to explore how deliberately integrating Indigenous, queer, and trans analysis can shake up top-down governance in river restoration, by theorizing a river model of justice. The argument unfolds as follows.

Chapter 1 draws on ethnographic evidence and Karen Barad's concept of apparatus as intra-action to argue that different water apparatuses and their associated imaginaries produced distinct, and sometimes conflicting, conceptual underflows: the more-than-human commons and hyporheic connection. I trace these two imaginaries' emergence and explore their potential to challenge top-down governance in Salmon Creek and the Scott Valley in California. I argue that these imaginaries, working in concert, can unsettle colonial tactics of management such as the doctrine of beneficial use.

Chapter 2, "Queer × Trans × Feminist Ecology: Toward a Field Praxis," explores how a queer-trans subjectivity modifies access to field studies in ecology. In the field sciences—ecology, conservation biology, hydrology, forestry, and geology, the "field" is usually located in rural or remote areas, in social milieu that are marked as heterosexual, white, and predominantly male. At worst, queer trans scientists, especially those who don't pass as white, are often in danger in these spaces. At best, out queer-trans scientists are welcomed awkwardly into normatively cishetero field sociality. Queer trans scientists might avoid interactions in single-gender dorms or open-air peeing and wetsuit changing, at best awkward and at worst dangerous. Instead, we might form relations through animals, plants, or the water, (by talking with or about them, as a shared form of recognition and love with nonqueer people) or with them directly, in solitary field encounters. I explore the affective dimensions of queer trans field work through José Esteban Muñoz's writing on "the problems of belonging in alterity."[69] Thinking with Muñoz and Eve Kosofsky Sedgwick,[70] I consider how queer x trans x feminist science practices can resist the unmarked whiteness and the colonizing impulses of queer and trans and create more breathing room for Black, Indigenous, POC, queer, trans, and other outliers in environmental science.

Chapter 3, "The Watershed Body: Trans and Queer Moves in Beaver Collaboration," considers trans potentials of multispecies collaborations along rivers, theorizing a transspecies watershed body cocreated by humans and beavers. Thinking with trans scholars of embodiment and becoming and transgression, I present beavers as stochastic actors against human river control projects.[71] In the Pacific Northwest, some ecosystem managers and scientists are turning to beavers as partners in restoring hydro-ecological function to damaged salmon streams. Drawing on interviews conducted on beaver relocation trips and at beaver management conferences, I propose that a trans figure of a watershed body can help humans emulate beavers' reshaping of water and land, rather than trapping beavers into service as "ecosystem engineers." I propose that Fred Moten and Stefano Harney's notion of improvisation within the undercommons could be a score for human-beaver relation that resists instrumentalizing beavers or their riverine worlds.[72] Bringing trans embodiment and improvisation within the undercommons to the more-than-human commons provides a way for engineers to subvert nature-control paradigms that are deeply embedded in their discipline.

Chapter 4, "Unchartable Grief: Scientists Grapple with Extinction Politics," takes up queer performance of grief and mourning as a model by which ecologists studying vanishing species can mourn their study subjects in public. Through fieldwork, we ecologists come face-to-face with organisms and landscapes that have been damaged, fractured, or pushed to extinction, and we often form intimate associations with these ecosystems and organisms at our field sites. But "straight" ecological scientific norms imply that we should deny such affective relations by suppressing queer feelings of kinship with other species. I theorize a queer politics for field ecologists that takes up Elizabeth Freeman's analysis of queer habitus as kinship.[73] As evidence, I describe encountering dying salmon and failing to include my grief for those fish in an ecology paper. I also consider ecologists' public expressions of love for their study species in a twitter #cuteoff and describe a participatory performance I staged where scientists confessed their troubles about extinction to a person dressed in pink and silver and wearing a papier-mâché salmon head. What would happen if scientists joined queer, Black, Indigenous, and environmental justice activists in public mourning or, inspired by them, began to perform multispecies politics in the public eye? More than enhancing science communication, such affective moves could (re)animate political movements for biodiversity, climate justice, and water protection.

Chapter 5, "With and for the Multitude: Cruising the Waterfront with José Esteban Muñoz," uses queer ecopoetics methods to attune to queer ecologies along an urban river. In a posthumously published piece, José Muñoz notes, "Queer thought is . . . about casting a picture of arduous modes of relationality that persist in the world despite stratifying demarcations and taxonomies of being."[74] This chapter takes up this sense of queerness in order to examine various practices of relating among humans and nonhuman entities that complicate readings of urban waterfronts as either industrial or natural. I argue that the brown commons could be the ground from which a queer mode of environmental justice politics springs. As environmental justice's queer ecological desire, the brown commons embodies utopian thinking for, with, and on the part of the damaged field of polluted and degraded landscapes.

In the epilogue, I recapitulate ways that underflows methods and queer-trans-feminist practices can transfigure riverine science by describing a new collaboration with Karuk cultural practitioners Lisa Morehead-Hillman and Leaf Hillman, to develop a river model of justice for the Klamath River. I explore how grounded relationality can spring from such collaborations and reconsider how queer aesthetic practices can work to unsettle rivers and their sciences.

While taking as its case sciences bound up in watershed and salmon governance, this book is mostly not about governance, politics, water use, or history, though I have written about these dimensions elsewhere and gloss that context in the chapters.[75] Rather it is about the affective intensification (following Jasbir Puar) that happens in the field and that comes into those social spaces because of the field and that shapes conversations and feelings.[76] This book is about intimacies that traverse back and forth from the field to the indoor rooms where most scientific data analysis, writing, and debate takes place. It is about what it means to be in the field in a body that is marked at different times as queer, as trans, as white settler, as a scientist, and as not enough of a scientist (but rather a social scientist, performance artist, or activist). That is, this book is about doing science with a body, orientation, and praxis that variously passes inside and remains beyond the boundaries of mainstream social relations in settler-colonial US society. Across different field sites and research collaborations, I ask what difference these modes of being and of being-seen make—for science and governance in ecology and hydrology and for a politics of solidarity with communities in struggle for water and environmental justice.

Underflows invites scientists who want to integrate river science practice more deeply with cultural and environmental justice concerns to use underflows methods to think through how their fields approach race, gender, settler colonialism, and ethics. I hope to inspire you, scientists, to articulate those stakes and commitments within your academic and management circles. I especially invite outlier scientists to share your fabulous and thickly storied approaches to doing science and making more just ecologies.

Thinking beyond academic thought and river pragmatics, I hope this book will help queers and trans people see science as a sexy and radical pursuit and to see the pursuit of relations with rivers through close observation and reciprocity as meaningful. When party queers looking fabulous along the river put on a snorkel and see the fish in the depths for the first time, when straight scientists affirm their fishy loves in public, and when planners design waterfront parks for queer lives as well as nuclear families, I see queer trans ecologies coming into view. Queer trans lives matter for rivers.

Underflow 1

Hyporheic

The hyporheic zone in hydrology is the hidden, interstitial zone beneath a streambed, a zone of rocks and sand saturated with water. It is a place of fugitivity and transformation—bugs and fishes go there to escape drought; microbes break down and take up nutrients. Water wells up and down, in whorls and seepages that move slower as the water above the streambed seeps away, then dries up. Still, the hyporheic moves, barely, as groundwater flows in from springs or pathways through the soil. Willows and other riparian plants tap these springs with their roots and chase the retreating water table. These soil channels and gravel paths carry different waters, some rich in carbon and nitrogen from decomposing algae blooms, others nutrient- and carbon-poor from passing through thin soils. In ecohydrology, these water flows and carbon and nutrient cycles drive a stream's metabolism—the balance between primary production (photosynthesis) and respiration (microbial decomposition). Respiration consumes oxygen that diffuses through the water surface and mixes in as water tumbles over rocks or that algae release during photosynthesis. Oxygen levels rise during the day as algae photosynthesize and fall after dark. This likening of a stream to a breathing body has symbolic potential to set up an affective resonance with our own human bodies, if we let it.

Underflow, in this sense of the hyporheic, influences and is influenced by other flows, which hydrologists term inflow, outflow, throughflow, overbank flow, and shallow subsurface flow. In early California groundwater law, underflow was used in a highly speculative way before hydrologists understood dynamics of groundwater and surface exchange.[1] These days, computer models allow hydrologists to track

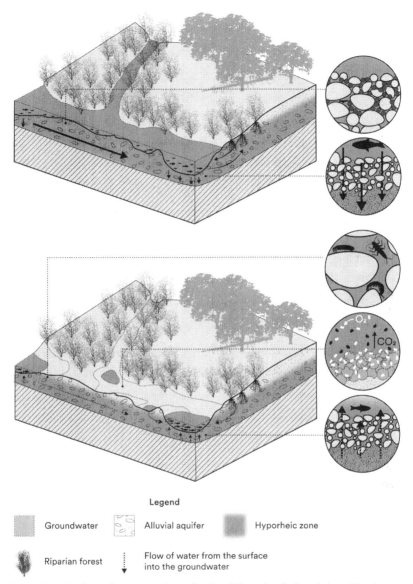

Legend: Groundwater, Alluvial aquifer, Hyporheic zone, Riparian forest, Flow of water from the surface into the groundwater

Figure I1.1. The hyporheic zone, from the Greek *hypo* (under) and *rheos* (flow) (Top) During winter and spring, streams flow aboveground, forming riffles in shallows and pools in between. Water from the stream flows downward, recharging groundwater. Salmon dig redds at the head of pools; young alevins hide in the gravel until they can outswim predators. (Bottom) During summer, water tables drop, leaving pools disconnected by dry riffles. Aquatic insects migrate down into the hyporheic zone and take refuge. Groundwater flows upward into the pools and mixes with subsurface flow moving downstream. Microbial slime coats rocks, consuming oxygen as it decomposes. Figure by Morgan Southall.

and predict when and how water will move through a stream. These models often include groundwater and surface water, but to incorporate the hyporheic zone adds yet more complexity. To build a model, hydrologists first define a region of interest—which could be a two-hundred-meter reach of a small tributary or could be an entire watershed. These models treat the surface of a river as one box and the groundwater from the streambed to the bottom of the aquifer as another. Modelers then direct the model to subtract some volume of water from the subsurface and add it to the surface to represent upwelling or subtract it to represent downwelling. Some small-scale models include the hyporheic zone its own "pool of water," often to simulate changes in temperature that occur as water wells up and down through the streambed's heat-trapping alluvium.

Staying ankle-deep in the stream's literal underflows, say in a pit I once dug to unearth a temperature sensor buried in gravel during a flood, let us consider a closely related underflow. This is a conceptual underflow within river management: how do ecosystems respond differently to groundwater or hyporheic flows than to surface water alone? In Mediterranean climates like California's, almost all precipitation falls in a few winter storms, then is stored as snowpack or as groundwater; without those reservoirs, streams go from low to dry by the end of the summer drought. In such intermittent streams, sustaining and recharging groundwater can keep streams flowing longer into the summer. As a boon for salmon, the groundwater that seeps into streams is often also cooler than water flowing over the surface.

Mainstream river science and management has focused almost exclusively on changing how much of a river's surface flow people divert, and when. With climate change increasing stream temperatures beyond the tolerance of salmon and other sensitive species, managers in California now look to underflows' cooling power. When water flows down through the hyporheic zone and back up again, it cools by several degrees. This can happen when water ponded behind a beaver dam plunges down into the bed or in an intermittent stream when flow drops below the riffle crest. Although research on how hyporheic flows affect stream ecosystems goes back to the early 1990s, only in the last few years has water policy and management begun to respond to this science. Water law in the arid US West has long considered surface water and groundwater as separate domains and has regulated land use with little consideration for how agriculture, forestry, mining, or urbanization affect water flow downstream.[2] Now, science has accepted that different streams have different dependence on groundwater. Underflows are thus critical to climate change adaptation, to store snow that melts earlier or falls as rain.[3]

My ecological research on intermittent stream dynamics on Salmon Creek has helped to push this field scientifically and to move this scientific discourse from academic journals into agency and grassroots river governance.[4] Like the hyporheic zone, small intermittent streams were long dismissed in studies of potential salmon habitat as outside the norm; only after citizen scientists and local consultants did their own studies was Salmon Creek included in federal coho salmon recovery plans. We salmon ecologists know now that the fishes move around, they find cool upwelling, they might smell groundwater because of its chemical composition, and they suffocate and die when pools are disconnected for more than a few weeks. We know that every small increment of water that stays in these streams increases the chance that young salmon and trout will survive through the summer drought. Now comes the work to put that knowledge into water management, by holding back the water that rushes down streams during floods, so it can recharge aquifers. This will take many strategies—reconnecting floodplains, using agricultural ditches to flood fields, bringing back beavers, and restoring high mountain meadows.

1

Thinking with Salmon about Water

More-than-Human Commons and Hyporheic Imaginaries

Introduction: Imaginaries Coconstituted with Apparatus

Let's return to the Scott River, to where it cuts through an arrow-straight gap in mine tailings: towering mounds of head-sized rounded rocks, pitched to a precipitous angle of repose. Sugar Creek flows into the Scott. Before it does, it runs through more tailings lingering in two long pools, fed by cold underflows. Now, in August, the river's bed is dry. Heat scorches the treeless rockscape, but down by these beaver ponds, tall willows and alder throw cool shadows. Damp air rises from the leaves and still water. Downstream, the river sinks into the tailings and disappears.

These tailings were created back in the 1920s when gold dredges upended the floodplain into a long stretch of parallel hills. The beaver dams were created too—not at first by beavers but the Scott River Watershed Council, a local conservation group. In 2015, volunteers wove willow branches through rows of posts, which were driven into the streambed perpendicular to its flow. They hoped these structures, called "beaver dam analogues," would lure actual beavers back onto the landscape. The support posts worked: beavers shored up the human-built dams, creating those pools, which also raised water tables far out into the floodplain.

These "beaver dam analogues" are an experiment in ecological reconciliation. They are one strategy in a long campaign to reshape social-ecological relations in this high desert. Hay and alfalfa production here depends on irrigation through the dry summer, using water either diverted from tributaries or pumped from a vast alluvial aquifer: deep sand and gravel that lie beneath the valley floor. Tens of thousands of Chinook (*Oncorhynchus tshawytscha*) and coho (*O. kitsuch*) salmon and steelhead trout (*O. mykiss*) once spawned in channels branching through a vast beaver meadow. By the late twentieth century, mining, levees, and farming had all but destroyed that habitat, and these changes pushed all three species toward extinction.

In some sense, the Scott River Watershed Council's beaver dam analogue project began in 2010, when the California Department of Fish and Wildlife (CDFW) reinterpreted the Fish and Game Code, telling ranchers that all water withdrawals in the Scott Valley would henceforth require streambed alteration permits.[1] The code is typically used to regulate irrigation dams (which can block migrating fish) or dredge mining (which destroys the gravel redds that shelter salmon eggs).[2] CDFW argued that water diversions and groundwater pumping dry up streams, killing fish and wildlife, and thus require its review. The Siskiyou County Farm Bureau, representing agricultural interests, sued CDFW for breach of authority, arguing that the regulation kept irrigators from exercising water rights. In 2012, the Siskiyou County Superior Court sided with the Farm Bureau and enjoined CDFW from requiring notification.[3] However, the Third Appellate Court reversed that decision, arguing that gold rush–era legislators who drafted the section in question did intend "diversion" to include groundwater pumping and would have considered dewatering a stream reach an "alteration" of the streambed. Agricultural water users are now required to file a permit application, to monitor impacts of groundwater pumping and freshwater diversion on fishes, and to identify and mitigate for fish kills.[4]

The decision affirmed the Quartz Valley Indian Community's contention that irrigation diversions harmed salmon and strengthened the Tribe's legal standing to manage groundwater and streamflow on Shackleford Creek, which flows through its reservation. Around the same time, in the depths of drought, the California legislature passed the Sustainable Groundwater Management Act, forcing changes in who can participate in groundwater governance. Quartz Valley now has a seat on the new Groundwater Sustainability Agency, which is facilitated by an outside expert. Such reinterpretation of legal and regulatory responsibilities can shift water use practices and reshape settler-Native politics and alliances.

In places like the Scott Valley, where water is scarce or unreliable, water shapes settler and Native environmental imaginaries: distinct—and often incommensurate—ways that ranchers, fishermen, and rural telecommuters value water and understand its movements from aquifer to stream to spigot. Through memory, measurement, and record keeping, people make sense of water's temporalities and circulation and develop norms and regulations to protect the uses they value most.[5] Whereas settler land ethics and governance paradigms emphasize stewardship of private property for resource extraction, Indigenous legal-governance paradigms, such as the Paulatuuq system that Zoe Todd describes in the Canadian Arctic, arise from philosophy and praxis of reciprocal responsibility. Similarly to how Paulatuuq people "assert legal-governance paradigms and Indigenous legal order to protect the well-being of fishes in the face of complex colonial and environmental challenges,"[6] the Quartz Valley Indian Community and other Klamath Basin tribes care for fish with science and ceremonial protocol and by using treaty law in dam removal and water management forums. I first describe the distinct imaginaries that shape Native, rancher, and NGO management, then explore the potential for alliances built on reciprocal relations to reshape water and salmon governance in the water-scarce West.

The term *imaginaries*, as I use it, describes a felt sense of connection to other people, waters, and species, which people develop as they work in and seek to protect their home landscapes.[7] Grassroots environmental movements often catalyze this sense of connection, as people respond to threats to the places they care for and rely on. Imaginaries are more than merely discursive or narrative—water imaginaries are rooted in place-based cultural and political struggles, and they arise as people use and care for water.[8] As water imaginaries travel, via media and social networks, from local environmental actions to global forums, they materialize "vast networks of interlinking, discursive themes, images, motifs, and narrative forms that are publicly available within a culture at any one time, and articulate its psychic and social dimensions."[9]

Building on geohumanities' theorization of relations with the lithosphere, islands, deltas, and contaminated landscapes, I take up the question "What is water, in this place and at this time? Where does this idea come from, and what does this idea do?"[10] I focus on imaginaries that circulate in local and regional governance of rivers and aquatic species.[11]

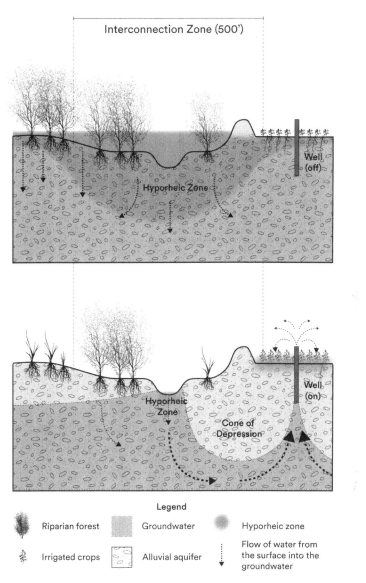

Figure 1.1. Water as relation and imaginary: floodplain cutaway
(Top) During the wet season, streams spread out onto floodplains. Water percolates down and recharges groundwater. The hyporheic zone expands, and ground and surface water mix. (Bottom) During the dry summer, the water table falls below the level of the streambed. Pumping pulls water away from the stream into a "cone of depression." In alluvial aquifers, wells far from the stream indirectly affect stream water levels; however, current policy in the Scott Valley defines wells more than five hundred feet from the stream as not connected hydrologically. Figure by Morgan Southall.

Some of these imaginaries prop up taken-for-granted settler categories, such as the idea of a sharp boundary between ground and surface water that is inscribed in California water law.[12] To understand how different people's imaginaries of belonging within a waterscape circulate and change, I focus on their everyday relations to water and on the apparatuses that enable those relations. I do this in two California watersheds where people seek to transform residential and agricultural water use so that salmon can return from the brink of extinction: the Scott Valley, in Siskiyou County, and Salmon Creek, in Sonoma County.

Apparatus, as I use it, takes up Karen Barad's sense: a contingent and lively array of singularities that "intra-act," producing phenomena.[13] For example, infrastructures, ecosystems, climate patterns, political-economic forces, and people's everyday ways of using water are all elements of a household's water apparatus, which intra-act to produce a place- and time-specific flow and an understanding of that flow. This concept of intra-action highlights reciprocity among humans and nonhumans as it refuses Descartes' stark separation between a human, thinking subject and a scientific, political-economic, or historical object. The apparatus and intra-action are central to Karen Barad's theory of agential realism, which explores how the specifics of scientific practice matter for scientific results and for the politics and policies that follow from them.[14] Science as a Baradian phenomenon includes race, class, gender, caste, politics, and injustice—rather than exiling these considerations to other disciplines. Since tribes, NGOs, and agriculture associations often rely on science to adjudicate disputes that are rooted in ethics, values, conflicting legal doctrines, and historical injustice, this wider domain for science is necessary. My analysis of imaginaries, informed by an understanding of water systems as Baradian apparatuses offers a way through the impasse of competing scientific studies and models, by articulating the imaginaries that underlie these values. When imaginaries conflict, competing models may entrench dissensus,[15] but when values and imaginaries enter explicitly, water governance can make space for subjugated knowledges, for Indigenous law, and for ethics that then reshape imaginaries. I think, specifically, with Barad here because their figure of the apparatus is especially generative for demonstrating how river sciences' questions and goals are coproduced by imaginaries, scientists, their devices and methods, and the actions of waters, minerals, and species.[16]

The analytical power of the apparatus extends to two influential arguments in the political ecology of water. For Jamie Linton, Horton's abstracted vision of purified H_2O—that classic image of water evaporating from oceans, condensing in clouds, falling as rain and snow, and flowing down through rivers, dams, and aqueducts—obscures social and political forces that shaped the emerging science of fluid mechanics. Linton argues that dam and aqueduct infrastructures first turn water from a local, animate relation into "an abstract, isomorphic, measurable quantity" and then remove it from ecocultural contexts.[17] Linton's analysis of river science and infrastructure as coconstituted with social and political forces reveals axes of power that was consolidated within the discipline of hydrology and the Bureau of Reclamation. For example, the idea of diverting water from the Colorado and Sacramento Rivers to the fields of California implicitly rested on an imaginary of water that can be exchanged without risk of ecological, social, or spiritual harm.

In *The Organic Machine*, Richard White describes ongoing Native governance, engineering, burning, and fish harvest in the Columbia River as always contesting and reworking settler extractive policies. This contestation ultimately produces a hybrid of "organic" and "machinic" elements that fundamentally reshapes Native and settler societies.[18] Thinking this organic machine as an apparatus that is material and discursive reveals the river as a lively phenomenon, beyond human control. Its animacy emerges from the "mixture" of dams, fishes, fishers, and other workers who "knew the river through the work the river demanded of them" and the river's own work—work in the physics sense of the exertion of force overcoming resistance in its rush downstream.[19] White's water body is a kind of cyborg body in which the machinic pieces cannot be separated from the living ones.[20] Like White, I am interested in how people come to know the liveliness of a stream and its waters through labor—mucking out springs, measuring stream temperature, weaving beaver dam analogues, counting salmon, and mapping flows of water above and underground.

Where human and ecological needs for water spur conflict, distinct apparatuses produce imaginaries. Two of these I call *hyporheic connection* and *multispecies commons*. Along the Scott River, reading relations through alfalfa irrigation apparatuses revealed hyporheic imaginaries that emerged during debates among tribes, ranchers, and conservationists on how to recharge aquifers and make dry rivers flow again. Hyporheic imaginaries prompted ranchers and salmon advocates to enlist scientists to study whether beaver dams could keep streams flowing longer by recharging groundwater. In materializing waterscapes as groundwater–surface water connection, hyporheic imaginaries disrupt a foundational tenet of California water law, which governs ground and surface water separately.

Multispecies commons imaginaries think waterscapes as more-than-human relation, challenging settler regulatory practice that often dries up fish habitat in order to keep water flowing to cities and farms. Along Salmon Creek, reading relations through household water apparatuses revealed multispecies commons imaginaries that emerged as residents watched near-extinct coho salmon return to local streams. Some of these residents came to see the stream as animate in its circulation through the watershed's channels and life forms. These imaginaries prompted some residents to adapt their water infrastructure and reduce water use to save more for salmon. Where people hold both imaginaries, I see the greatest potential for collaboration and scientific innovation.

To think of a river as a multispecies commons (one imaginary) or to imagine hydrologic connectivity through mine tailings or beneath farm fields (a second imaginary) can start to transfigure water politics and relations among human actors. In Salmon Creek and the Scott River, climate change is already making rivers hotter and summers drier and is exacerbating past settler habitat degradation. Each imaginary comes with its world—of people, concern for cows or salmon or beavers, attitudes toward science and regulation, and politics. These imaginaries reveal matters of concern and care that animate conflicts over how to govern water in a given place and how to care for the land, water, and species.[21] Indeed, these worlds, and relations of care for them, are the Baradian apparatus that materializes distinct water phenomena.

Commons as Indigenous Erasure

In Salmon Creek, as in many settler communities, commons and commons imaginaries often erase Native land tenure and Native people's ongoing use and management of the land. After I first elaborated the multispecies commons in relation to Salmon Creek, Glen Coulthard's critiques of the commons made me question whether commons can do unsettling work. The lands that settlers make into private property or commons are, as Coulthard notes, Native lands, which are often unceded. On these lands, "native and customary relationships that emerge from Native connection to the land are integral to imagining new formations of private property."[22] Here, Coulthard gestures to commoning strategies of Native peoples, which persist unrecognized by settlers, who often look to Europe for models to deploy in North American settings. Asked whether commons could center Indigenous presence in radical struggles, Coulthard replied, "If we are committed to reclaiming the commons we are going to have to work critically to re-establish non-capitalist and decolonial social relations and legal traditions that have survived through generations of Indigenous communities."[23]

Here, I use the term *multispecies commons* and highlight this tension around erasure and potentials for solidarity in Salmon Creek, the Scott Valley, and beyond.

Approach

Staying close to material underflows is key to my method. I follow water's hidden courses through the subsurface, its physical upwelling and seepage and biochemical transformations en route from sky to stream to aquifer and its transits through plant, animal, and microbial bodies—especially bodies of salmon.[24] As White notes for the Columbia River, salmon habitats were cocreated through Indigenous human work—including burning, cultivation of riparian plants for food and fiber, and selective harvest—the "work" of beavers, trees, and the river itself. This work routed much of a stream's flow via complex underground paths. On both Salmon Creek and the Scott River, people experiment with different ways of modeling and measuring underflows, and they contest one another's imaginaries of its flow in public meetings and documents, where they try to balance agricultural and salmon-based livelihoods. These ethnographic and archival sources ground my analysis. I focus on the Quartz Valley Indian Community and Native and non-Native river restoration organizations as they interact with regional, state, and federal regulatory processes.[25] All of these entities rely on scientific studies—often funded by state and federal water and fish agencies—to direct water and salmon management. In both cases, diverse knowledge holders and practitioners negotiate goals, approaches, and sites for study. To practice feminist science in this context is, first, to understand that these negotiations of science and policy are always infused with history and power. My feminist science practice works to confront and dismantle injustice by working with tribes and grassroots groups, mindful of power differences that arise from this history. I channel research funds and academic labor to Native and place-based communities, and I amplify Indigenous science as one powerful underflow to white supremacist surface flows of water policy.[26]

Applying apparatus as an analytic to explore coconstituted imaginaries is a method of transgressive scientific collaborative inquiry that brings Indigenous science, Western science, and local knowledge into contact.[27] I designed a field interview method to formalize how feminist STS standpoint approach could yield stronger objectivity and catalyze action on streams.[28] I also facilitated four annual collaborative workshops where local residents and scientists hashed out how to increase streamflow and help salmon.[29] These data ground the Salmon Creek case and imaginaries. On the Scott River, I worked from legal opinions, groundwater study proposals, public comments, and other public records. In both cases, I apply the analytic of apparatus to search for attendant imaginaries of hyporheic connection and multispecies commons and to explore how people's encounters with salmon and beavers, and new scientific understandings of their dynamics, reshaped imaginaries.

Case Context: Surface and Underflows in California Law and Policy

California's megawaterworks move water from mountain rivers to fields and cities. Throughout the US West, such networks of dams and levees consolidated settler-colonial land tenure as their architects invoked Manifest Destiny rhetoric that white settlers would improve uncivilized landscapes by controlling unruly rivers, suppressing forest fires, and harvesting "wasted" trees, fishes, and game.[30] The hydrologic models and methods they developed dominate top-down water governance to this day; because dam and levee planners devalued aquatic ecosystems and lifeways connected to rivers and wetlands, salmon-based livelihoods now depend, precariously, on industrial hatcheries that are failing to compensate for habitat destruction.[31] By freeing farmers and municipal water companies from dependence on local streams and aquifers, this unprecedented river engineering project also created an artificial divide in water law and policy, between ground and surface waters that are hydrologically connected.

In the 2000s, two plans challenged orthodoxies embedded in the twentieth-century development of California's waterworks, that

1. large waterworks and centralized water management schemes can expand with little regard for ecological limits,
2. centralized waterworks can best protect California residents from droughts, floods, disease, and economic stagnation, and
3. groundwater can augment surface water diversions for irrigation and domestic use without considering how groundwater pumping affects streamflow or other wells.

The 2005 California Water Plan Update "[strove] to meet all future water demands—urban, agricultural, and environmental" and encouraged decentralization of some water governance processes through Integrated Regional Watershed Management Plans.[32] The 2014 Sustainable Groundwater Management Act mandated new groundwater sustainability agencies, which must devise groundwater management plans "that avoid significant and unreasonable adverse impacts on beneficial uses of interconnected surface water."[33] Local water management, which is often exclusionary, is now interfacing with new streamflow rules and climate adaptation ideas through state-facilitated collaborative processes. Within this changing regulatory landscape, collaborative, transdisciplinary science that supports local self-regulation and governance can produce novel understandings of riverine ecosystem function, through imaginaries of multispecies commons and hyporheic connection. Rural landowners often challenge or ignore top-down water regulation, favoring self-regulation and voluntary measures. Vast numbers of private wells and streams are nominally governed by state rules but escape de facto regulation.[34] Salmon Creek and the Scott Valley exemplify this situation. I now look more closely at the imaginaries that emerged in Salmon Creek when a local coalition trying to avert imminent salmon extinction started investigating local water flows.

Salmon Creek: Household Apparatuses, Scarcity, and Imaginaries of the Multispecies Commons

A few hours north of San Francisco, vineyards blanket the Russian River valley, watered by black plastic drip irrigation lines that carry water pumped from the river. A few old apple orchards interrupt the carpet of vines; they're not irrigated but also not as lucrative as grapes. Just over the ridge to the south, by Salmon Creek, people rely on smaller, unreliable water sources. The small and troubled Bodega Water Company supplies water to sixty-three connections, including the church made famous in Hitchcock's *The Birds*, one bar, a country store featuring a foam rubber Hitchcock figure, and an antique shop/yoga studio. Bodega Water customers described a purple-brown tinge and metallic taste in water during the summer, service interruptions, and high water rates. Descendants of Italian settlers run dairy cows on the grassy slopes above Salmon Creek, since the creek's trickling flow won't support irrigation; it dries to a series of isolated pools by August in most years, shutting down permitted and illicit pumps. The remaining households, businesses, and dairy farms depend on rainfall (stored in ponds or tanks) or springs and wells that tap shallow aquifers. In a Baradian sense, the water apparatuses here include these diverse water sources: pipes, pumps, and tanks; monitoring practices; hydrological models; regulations; and the fishes, willows, and rare shrimp that live in the stream. These apparatuses materialize different phenomena under different water use practices.

Up on the redwood-cloaked ridges, differences in geology and land use produce different quantities and qualities of water. Beginning in the 1970s and accelerating in the '90s, ranchers cut off the redwoods and sold lots to migrants from cities, where "we didn't have to think about water."[35] New residents rely on springs or deep wells that tap sandstone lenses on the ridgetops or draw water from fractures in metamorphic rock. On a few larger properties—ranches and former communes—people tap larger springs for drinking and to water gardens and livestock. Only a handful of households described having such productive, high-quality wells that they never had to worry about water.[36] Many ridgetop residents' well levels dropped during the drought, and in a few cases, springs or wells dried up for the first time. These users often worried about water and sought to secure backup supplies as a hedge against an unpredictable climate (or forgetting to turn off the hose). Many opposed new development, fearing it would dry up their wells.

Before the drought began in 2011, Salmon Creek's water apparatuses were all very different from a city apparatus, where users turn on the tap and never think that water won't flow out. Many Salmon Creek residents wished for a backup supply, and some built rain catchment tanks. But since most ridge dwellers rarely visited the creek, they didn't connect their experiences with scarce or unpredictable water to stream conditions and salmon survival. That changed in the early 2000s, when some local residents banded together to save the dwindling salmon runs.

Salmon Decline Shifts Water Imaginaries

Along California's Central Coast, coho salmon decline began in earnest in the 1950s. As elsewhere, this decline was death by a thousand cuts: dams, logging, beaver removal, levees, and agriculture all changed where and how the stream flowed. By the mid-1990s, state biologists found no coho in their surveys of Salmon Creek and declared them locally extinct. The National Marine Fisheries Service listed the population under the Endangered Species Act in 2004. Their recovery plans named valley bottom lands across the central California coast as potential critical habitat, setting up conflict with rural residents, ranchers, and vineyard owners. Recovery, the agency asserted, depends on how all of these people manage land and use water.

While these plans were still under discussion, in 2001, biologist Michael Fawcett noticed that the last remaining year-class of Russian River coho were about to perish in drying tributaries; biologists brought them to a conservation hatchery.[37] These fishes and their descendants now live out their lives in freshwater tanks. Biologists released captive-bred coho fry into tributaries each spring beginning in 2004; their descendants have started to rebuild the run.

In 2010, a few Salmon Creek residents formed a watershed council, which began holding community education events and conducted scientific studies of salmon habitat and streamflow conditions. These studies found some good spawning and rearing habitat but little water in the summer; some reaches dried up completely.[38] When the pumps in Bodega's municipal wells, dug next to Salmon Creek, turned on at night, water levels in the creek dropped in sync. These findings supported a proposal for federal salmon recovery funding to install rain tanks to reduce pumping, thereby keeping more water in the stream for salmon. To be eligible for a subsidized rain tank, residents agreed to use only rainwater for irrigation. The grant funded 90 percent of the cost of tanks, which ranged from 9,000 to 80,000 gallons (34,000 to 303,000 L), plus four years of salmon monitoring. That year, local rain gauges measured the lowest rainfall in decades, and one resident watched his spring dry up for the first time in the thirty-five years he had lived there. Private water trucks rumbled back and forth, bringing Russian River water to residents whose wells had gone dry, at $150 per 3,000-gallon (~10,000 L) truck.

One rainy spring day that same year, a few biologists and locals carried three hundred adult coho salmon from a truck to the edge of the water near the mouth of Salmon Creek and released them.[39] The fish slithered and splashed upstream, then spawned. These were surplus adults from the Russian River captive breeding program, not needed to continue the hatchery lines. By summer, several tributaries had nearly gone dry, and dissolved oxygen in the small pools that remained dropped toward zero. State biologists collected the finger-length fry in nets and took them back to the hatchery. Once the rains began in November, biologists returned the fish to the stream, where they lived for a few more months before (presumably) swimming out into the ocean.[40]

Each winter since then, a few humans have released hatchery salmon in the estuary to spawn where they will. This slight change in material/izing water conditions—a few hundred fish that swam upstream, spawned, and died—transformed the social and material interactions of the watershed's human residents. Carl Anderson, a longtime resident who lives just downstream from Bodega, described the return of the salmon in cultural terms: "That tree line is Salmon Creek. It's three miles out there, and you probably can't kayak it because of the trees that go across. There are people that seem to remember that there were a lot of fish in this stream at one time, and you could go spear them after school. And we see now, there's coho in there, and otters, and turtles. It's really good to see." The release of coho salmon from a conservation hatchery was a precipitating event. It took science, debate, and community meetings, in addition to gossip, weather, and fish migration and survival patterns, to crystallize a precipitate of new understandings and new actions. In a

Baradian sense, new phenomena emerged from different configurations and commitments of human actors, who never act alone.

Imaginaries and Apparatuses Shift in Response to Salmon's Possible Return

Economic constraints, a lived experience of scarcity, and a desire for salmon's return from the brink of extinction are fostering a multispecies commons imaginary.[41] These imaginaries emerge coconstituted with a shift in water use practices and build momentum for further restoration work among residents and agency scientists, and they are characterized by

1. an awareness of water supply apparatuses as place-based entanglements among people, their wells and springs, and fishes, trees, grasses, raccoons, and other species who also depend on local waters. This awareness arises from their lived experience of scarcity—heightened during the 2011–15 drought—that people extend to nonhuman creatures. Everyday practices like flushing the toilet or bringing water to the horses resonate with significance for tiny fry rearing in the tributaries.
2. a detailed local knowledge of one's own water source, of neighbors' wells and water use practices, and of local hydro-ecological cycles. People know the vagaries and quirks of their water sources, pumps, pipes, and tanks; conserve water so as to not run out; and gossip about less-frugal neighbors' watering practices. "Did you know she waters her lawn?" "No! Someone should talk to her."
3. the conviction that local self-regulation is preferable to outside regulation. People believe that their close attention to the rain and their wells makes them the best stewards of the water, and they don't want county or state regulations to limit or even measure their water. "I think that collaborative efforts work better, but until people get educated—people are angry when they're made to do things."
4. a desire to reconfigure household or neighborhood water infrastructure to increase summer streamflow and benefit salmon. Depending on where in the watershed they lived and attitudes toward outside regulation, people proposed building rain tanks, conserving water, thinning forests to reduce evapotranspiration, and recharging winter rains.

Most people said that Salmon Creek and its watershed and groundwater should be able to support farms, residential development, and salmon (and sometimes other riverine species). This is the imaginary that I call the multispecies commons; this imaginary is material-discursive and a part of the apparatuses. Though their politics and socioeconomic status differed, all the people who expressed a multispecies commons imaginary had several things in common. First, they understood climate as intimately tied to water supply, explaining that because of the shallow aquifer, more winter rainfall means that wells and springs will stay wet longer into the rainless summers. Most maintained a rain gauge and could recall annual rainfall for specific years; most also tracked the flow of their spring or water levels in their well or rain tank. This individual climate knowledge became collective when neighbors talked rain at the post office or bar. Second, residents believed that groundwater was a common pool resource,[42] and that others' use could impact their own supply. Many blamed neighbors, cattle ranchers, or grape growers for water waste, but few thought they themselves wasted water. Third, people often saw water as a connective fluid that tied together people, other animals, and plants. But they didn't often articulate how ground and surface waters flowed into one another—they lacked an explicit imaginary of hyporheic connection.

Carl Anderson's water for drinking, washing, and irrigating a greenhouse comes from the Bodega Water System. A neighbor once pressured him into serving on the Bodega Water Company board, where he saw the system's vulnerabilities firsthand. These vulnerabilities coconstitute the apparatus—municipal wells that dry up the stream; magnesium-rich bedrock that turns water dark and metallic tasting during summer; the aging network of PVC pipes; and the system's small size and high maintenance costs, which result in some of the highest water rates in the state. Also part of apparatus, El Niño storms recharge aquifers but also trigger floods that have breached supply pipes; wildfires have threatened the town in recent years, when creek levels were too low for firefighters' pumps. Recognizing that rain tanks reduced all of these vulnerabilities, Carl quickly signed up for the rainwater tank subsidy program.

Carl used a stick as we walked down to his rain tank. There, the stick functioned as a measurement device: when he tapped the stick against the polyethylene tank, it thudded below the water line and rang hollow above it. He estimated the rainwater would last until the first fall rains, but if he ran low, he'd stop watering the tomatoes. Standing at the rain tank, I asked my standard interview question, "Who do you think should decide how much water should be withdrawn from Salmon Creek?" It was his response that first suggested the concept of a multispecies commons to me: "Who's going to provide a habitat for the fish and the bobcats and the deer and the coyotes and the raccoons that go down to the creek to drink? You can hear them down there. Do they have a right to clean water? I happen to think that we all live here together as a living network.... The creek should be preserved for the benefit of all living people—all living beings, as well as to humans. If that means a regulation of consumption, then maybe we need to self-regulate." Although he chafed at government water regulation, Carl did think the county should block new development. And he worried that individuals might not self-regulate, leaving too little water for other species. By talking through these governance quandaries next to the rain tank and in sight of the rain tank, he conceived of and opened up to an imaginary of commons governance that benefited not just human residents, but nonhuman ones too.

Others didn't mention salmon or stream habitats as motivation to conserve water. A Portuguese bed-and-breakfast owner who split his time between San Francisco and Salmon Creek carefully stewarded water level in his rain tank to sustain a beautiful, low-water garden. Following common practice back in Portugal, he planted figs, olives, and ornamental plants that needed no irrigation, watering only vegetables. He knew intimately how much water he used in the garden, and he said that having and monitoring the tank had made him much more conscious of his water use and prompted him to conserve more actively: "I only have so much water available, forty-six thousand gallons, so I plan accordingly.... I tend to plant things like potatoes, garlic, onions early on so that when it stops raining I have already harvested.... In case of an emergency, if I have some plants, I wouldn't let them die. I will use the Bodega Water Company water. But this will be the last resort." However, he hadn't known about salmon's dependence on local streamflow until the local conservation district approached him about participating in the rain tank project.

In contrast to Carl and the B&B owner, a few Salmon Creek residents did describe water's subterranean flows as connected to streams. Ranchers referenced scientists' descriptions of wells as many straws that, collectively, sucked the stream dry. Ridgetop residents recalled the rainwater recharge slogan "Slow it, spread it, sink it" and were interested in building rainwater recharge basins on the ridges.[43] This imaginary sometimes occurred without a sense of a multispecies commons. One retired ridgetop resident described the aquifer his well taps as a shallow bowl of sandstone perched atop an impervious metamorphic layer. "It's a mess—there's fingers and little separate depressions. The only water we get, basically, is what we get in winter. It's saved there in this bowl until we run out. There are places in the woods here that are seeps, where it's overflowing from these bowls year-round," he said. This imaginary of underground connection between a subsurface bowl and his own well does describe specific subsurface topography and flow paths. However, it is spatially limited at his property line (where the seeps are located). It does not connect to the salmon-bearing stream downhill.

Before I began my groundwater study, most ridgetop residents didn't think that their water use decreased streamflow. During field interviews, I showed them maps of springs and aquifer boundaries, sparking ideas about how aquifers feed springs, which in turn sustain summer streamflow. While looking at this map, a longtime resident who lives near the stream remarked that his seasonal pond might recharge water that ended up in his well. He explained, "Around the beginning of June or the end of June the well dries up. During the time when the water table is high and the earth is full of water, [the well] works just fine." Only some, like Carl Anderson, who lived near the stream, monitored how seasonal groundwater fluctuations in their wells tracked variations in streamflow. These people closely observed well level, spring flow, and rainfall, and developed nuanced conceptual models of water's movement from the sky, through the aquifer, to the stream.

Diane Masura, watershed council stalwart and retired tradeswoman whose hand-built house features her intricate tile work, keeps detailed records of rainfall and well level that she showed me during a 2012 interview; for years afterward, she would call or email me to discuss the latest data and how it changed her understanding of hyporheic connection. Diane moved to a small hillside parcel of second-growth redwood forest in the 1980s. Diane rarely spends time in Salmon Creek—she sees it from bridges and occasionally at a friend's place, and she volunteers to walk a short stream reach as part of a citizen science mapping project that I and the watershed council developed. Her everyday interactions with water take place at an old hand-dug, brick-lined well, which supplies her house and a small flower garden. At the well each month, she lowers a homemade device made from a knotted string tied to an empty plastic water bottle, then records the depth. The string is marked in increments—not feet or meters, but her own height. She plots these depths against rainfall totals in an old logbook. In one poststorm email exchange, she said that the well filled up but then the water level quickly dropped; I asked how long it took to rise again. She replied,

> Dear Cleo, You were right. The well went back down, but not all the way back down to 15.5 feet (back to 16.125). I am used to visualizing this cistern well like a sink. When the rain comes rapidly, the well fills up. When the rain slows down, it drains into the underground ground level. I think I've said that the baseline used to be about 28'. Now the baseline is about 20'. So it'll take a lot more rain to get the underground level up to 20' or higher. Keep your fingers crossed. Be well, Diane

Diane first imagines her well as a kind of holding tank, intimately connected to the aquifer. Her body is part of the water apparatus, which is labeled according to the phenomenon of her concern and care. She feels intimately connected to the aquifer herself through her water use, and she is reminded of this connection every time she converts her measurements from units of Diane Masura to feet. Through this apparatus, the phenomenon of groundwater emerges —as a dynamic flow connected to the rain above and the stream below. Diane also described political and economic factors that affect her well—logging; local neighbors' care for the protected grove; exurban residential and vineyard development that she worries will further deplete groundwater; and California's lack of groundwater regulation.

Diane's imaginary of multispecies commons with hyporheic connection contrasts sharply with imaginaries of many of her ridgetop neighbors, whom I also interviewed. These were relative newcomers, who said that pumping from the stream down in the valley harms salmon but their own well use does not. They were acutely aware that their groundwater was scarce but did not see that groundwater as connected to streamflow.

Residents who used springs rather than wells as their water source, on the contrary, often held both imaginaries (multispecies commons and hyporheic connection) and thought of them as material and discursive. They knew which streams those springs fed, and they saw those springs as coupled to the rain and shallow aquifer. During the 2011–15 drought, several people saw their springs go dry for the first time. Those springs often held frogs, salamanders, and insects, which people encountered when they maintained pipes or pumps or measured flow. This proximity—human infrastructure and aquatic habitat interpenetrating one another—is one condition that enables a multispecies commons imaginary to emerge. Residents' repeated visits to measure and monitor their water infrastructure increase the strength of intra-actions. This repetition reinforces this imaginary, and may spur action to protect that commons from threat.

Such were imaginaries in circulation in 2012, when I conducted initial field interviews in the Salmon Creek watershed. Over the succeeding years, deepening drought and continuing coho salmon recovery have sparked new science and management practices. These practices continue to reconfigure apparatuses and imaginaries. As the drought deepened, many residents worried that others' wells would dry up their household supply. But until I involved community members in my tracer studies of connections between the ridgetop aquifer and streams, few people's imaginaries included the idea that their own water use could affect flow in the creek. These interviews in the Salmon Creek watershed revealed that competing imaginaries (long framed by groundwater–surface water distinctions imposed by settler water law) can shift when people see salmon in streams and trace flows from aquifers through springs to the stream. The combination of multispecies commons and hyporheic imaginaries has transformative promise for how people manage their everyday water use.

Scott Valley: Agricultural Water Apparatuses and "Zone of Interconnection"

In contrast to Salmon Creek, where aquifers store too little water for irrigated agriculture, the Scott Valley's abundant groundwater, sustained by snowmelt, supports an agricultural economy centered on irrigated alfalfa and beef cattle. After tumbling down steep drainages in the

Shasta-Trinity alps, the Scott River tributaries pour into a broad floodplain valley and sink into the ground, recharging a vast aquifer, estimated at 400,000 acre-feet (490 million m³).[44] The 500,000-acre watershed is home to around eight thousand people, including Quartz Valley tribal members and people, mostly white, who work in ranching. Ranching consumes the majority of Scott Valley water. During the irrigation season, which runs from May to early October, most water in the Scott River and its tributaries flows into irrigation ditches that feed the sprinklers that water alfalfa and hayfields continuously, turning 30,000 acres in the bottom of the valley a brilliant green.[45] Upstream of the alfalfa fields in the Scott Valley, spanning the Sugar and French Creek confluences, dredge mining left thirteen kilometers of the Scott River straightened, buried in massive mounds of tailings, and shoved against the valley wall; this reach also goes dry almost every summer.[46]

In a Baradian sense, a water apparatus is constituted by intra-action: among tributaries, groundwater aquifers, salmon and their habitat needs, irrigation ditches, deep agricultural and shallow residential wells, and past mining. Also intra-acting in this apparatus are tribal rights,[47] legal codes that specify who gets water and how much must stay in the stream, the institutions charged with enforcing these laws, and the science that Native nations, farmers, and environmentalists all do. Whereas in Salmon Creek, the precipitating event reconfiguring this apparatus was salmon reintroduction, in the Scott Valley, a different riverine animal—the beaver—shook up imaginaries of water flow, by adding its own infrastructure into water apparatus. This apparatus intra-acts to create imaginaries that center on one of two species—salmon and cows—and the human livelihoods they sustain. Beavers have recently transfigured both of these imaginaries and created a space of mutual interest and dialogue within the conflicted space of Scott Valley groundwater governance.

The salmon-centric imaginary is held by Karuk, Yurok, Hoopa, Klamath, and Quartz Valley tribal members; commercial fishermen; and some environmental groups, such as the Mid-Klamath Watershed Council. They focus on repairing habitat damaged by mining and stream straightening, on restoring flows of cool groundwater to dehydrated streams, and on adapting to climate change by restoring headwater meadows and reintroducing cultural burning to the forests. This streamflow is critical to spawning and rearing fishes—endangered coho and Chinook salmon and threatened steelhead trout, as well as Pacific lamprey—which are key food and cultural resources for the Quartz Valley Indian Community and other Klamath River tribes.[48] As salmon swim up the Scott River from the Klamath, the first major tributary they encounter is Shackleford Creek, which flows through Quartz Valley Indian Reservation. Coho salmon once spawned in large numbers in this prime habitat, but agricultural ditches and groundwater pumping now divert all of the creek's surface flow by July. Further up the valley, three other salmon-bearing streams face similar pressures; others once held salmon but now are too dry or degraded.

The cow-centric imaginary is held by ranchers, agriculture-focused institutions like the Siskiyou Resource Conservation District and irrigation districts, and other NGOs, such as the Scott River Watershed Council. They focus on securing water for irrigation, on market mechanisms to pay irrigators if they decide to dedicate some water to instream flows, and on maintaining ranching as a viable livelihood. Some ranchers work through the Conservation District and Watershed Council on salmon habitat restoration projects that remediate "legacy impacts," namely mining and levee building. A few ranchers are experimenting with using irrigation ditches to flood fields and recharge aquifers during the spring snowmelt season. Ranchers—few of whom are tribal members—irrigate alternately with ditches (collectively maintained infrastructures that carry water diverted from tributaries) and groundwater (pumped from the aquifer at water users' expense). Prior to 1950, most ranchers irrigated by diverting streams into ditches. After state regulators limited these diversions, ranchers turned to groundwater, which now predominates. Irrigation withdrawals increased 115 percent and irrigated land area increased 89 percent from 1953 to 2001, with most of the increased supply coming from groundwater.[49]

Within both the salmon- and cow-centric imaginaries, people agree that dredge mining destroyed important floodplain salmon-rearing habitat and blocked salmon from upstream reaches. As on many other California rivers, hydraulic mining and gold dredging scoured floodplains away and buried others with tailings, triggering floods that devastated farm fields and salmon habitats. In response, the US Army Corps of Engineers straightened the Scott's main stem, removed riparian forests, and built levees. This flood control network fixed the boundary of the river and expanded farmable land and pasture, but it also reduced aquifer recharge and caused the river to scour down into its bed, further lowering the water table. This simplified habitat supports less salmon spawning and mostly dries up every summer, leaving young salmon fry to die in shrinking pools.

Before 2010 or so, neither of these two imaginaries included beavers as agents in hydrological cycles despite ranchers, tribal members, and agency scientists all knowing that beavers lived in the Scott Valley. Many agency biologists believed that beaver dams blocked salmon spawning or movement; state trappers routinely killed "problem" beavers. Ranchers breached beaver dams if they blocked culverts or interfered with irrigation infrastructure. Some remembered that early trappers named the valley Beaver Valley—Stephen Meek, a fur trapper, described it as "'the best place for beaver I ever saw' . . . all one swamp, caused by the beaver dams" in 1836.[50] Some of these beavers hid out in backwaters, on the Scott and on rivers throughout the West, and are slowly returning to their former haunts.

In the mid-2000s, a team of NOAA scientists noticed that beaver dams on one reach of an Oregon stream were rebuilding the floodplain and that beaver ponds there teemed with steelhead trout. In other reaches, where the stream had carved a deep gully called an "incision trench," floods breached the beaver dams each year. The team, led by Michael Pollock, decided to try an experiment: if, in summer when flows were low, they reinforced beaver dams by driving posts into them, would the dams withstand floods the next spring? The scientists installed one hundred structures along the twenty-eight-kilometer reach.[51] After some tinkering by the scientists and the beavers, the stream transformed from a dry, wide gully to a braided stream through ponds; some of the dams did not persist, and others were abandoned, but on the whole, the approach seemed to work. Pollock's visit to the Scott in 2012 was part of what Daniel Sarna-Wojcicki and I have called a "beaver moment"—an upwelling of interest in beavers as stream restoration collaborators.[52] Pollock, like some of the people who have learned about these techniques at conferences or online, embraces beavers' unpredictable interventions into human engineering projects. Other settler scientists, wildlife managers, and ranchers tend to see beavers as beneficial where they can work for humans as "ecosystem engineers"—but only if humans can control their tree-chewing and ditch-blocking tendencies.[53] I will return to this beaver moment in the conclusion. Now, let us consider the imaginaries that circulated in the Scott and set the stage for beavers' transfigurative act.

The Flow of Water Imaginaries in the Scott Valley

By the 1990s, coho and Chinook salmon decline on the Klamath River spurred listing under the Endangered Species Act. This federal protection required that agencies make plans to improve streamflow and temperature, designate and restore critical habitat, plan for climate change, and study dam removal. An unprecedented fish kill in 2002 galvanized tribes and fishing groups around dam removal, after eighty thousand spawning salmon in the lower Klamath River died due to low water. Against its own scientists' recommendations, the federal government had released extra water to upstream irrigators, triggering an outbreak of the fish parasite Ich (*Ichthyophthirius mulifiliis*) in stagnant, warm water. It was against this backdrop that UC Davis groundwater scientists scoped a study of groundwater dynamics in the valley.

As in Salmon Creek, the Scott Valley imaginaries circulated in private meetings and public forums, deployed by different groups to affect river management: they were both material and discursive. These imaginaries are underflows to dominant groundwater and salmon discourse when they rework top-down approaches to science or contest management-as-usual. In 2008, during the groundwater study review, people who expressed different imaginaries both emphasized hyporheic or subsurface connection but differed in how they thought Scott Valley water should be considered legally. On one hand, ranchers and county officials justified maintaining large diversions from streams for agriculture under the doctrine of "beneficial use" (informally known as "use it or lose it"), which is enshrined in the prior appropriation system of western US water law. On the other hand, tribes, environmental NGOs, and fishermen's groups argued for returning diverted surface and interconnected groundwater to streams for salmon under the doctrine of "public trust," which has precedent in California water law via the Mono Lake Decision, which slowed Los Angeles's Metropolitan Water District's diversions and stabilized water levels in that lake.

The *hyporheic imaginary with beneficial use* holds that people should protect salmon populations insofar as doing so does not conflict with profit from agriculture. Rather than an imaginary of a multispecies commons that centers fishes and riverine creatures, this imaginary emphasizes settler stewardship over livestock and hay crops. It centers cows, but it accommodates wild species like deer and elk that are hunted for sport and food and sometimes graze in agricultural fields. Some ranchers include salmon in their imaginaries and describe how agricultural return flows create salmon habitat that fishes use—as a kind of mitigation for habitats dried up by diverting that water. Characteristics include

1. an awareness of water supply apparatuses that arises from an experience of abundant ground and ditch water and a perception that seasonal drying of Scott main stem and tributary reaches is either natural, happening even before European settlement, or a result of "legacy impacts" like mining, levees, and floods.
2. a detailed local knowledge of one's own water source, of neighbors' wells and water use practices, and of local hydro-ecological cycles. This knowledge circulates through management forums like watermaster associations, advisory committees, and day-to-day maintenance of shared infrastructure like ditches and diversion dams and gates. Unlike along Salmon Creek, people did not report a sense of water scarcity or ever running out of water.
3. the conviction that local self-regulation is preferable to outside regulation. Less formally, people believe that their close attention to the rain and their wells makes them the best stewards of the water, and they don't want county or state regulations to limit, or even measure, their water. This is very similar to Salmon Creek.
4. a desire to maintain existing water apparatuses, including ability to pump unlimited groundwater. There is an interest in voluntary measures like surface water leases but only if they are not tied to groundwater monitoring by public agencies and do not require making groundwater data public.

The *hyporheic imaginary with public trust* holds that groundwater and streamflow should be considered public trust resources that sustain fisheries, which are also public trusts.[54] Tribes have additional treaty rights and cultural relationships that are based on stewardship as relations within ecocultural landscapes, which share some characteristics with multispecies commons.[55] Characteristics of the public trust hyporheic imaginary include

1. an awareness of water supply apparatuses that is centered on hyporheic exchange between ground and surface waters. People perceive cumulative impacts of individual ranchers' diversions on the lower Scott River and the main stem Klamath and see groundwater as a source of coolness as well as flow. There is a sense of streamflow as diminished and sometimes totally exhausted by diversions for agricultural use.
2. a detailed local knowledge of salmon refugia: holes where spawners rest, tributaries where fry rear, and how streamflow decreases affect refugia. People have a temporal understanding of when fish should be in certain places and tell stories of salmon being widespread in the mid-twentieth century—after mining and levees but before groundwater irrigation.
3. the conviction that outside regulation of water adjudications, minimum flow standards, and illegal diversions is the only way to restore streamflow and salmon habitat. People have a sense of streamflow as not just a volume of water but a flow regime with a temporal rhythm that sustains riverine processes. For example, seasonal flooding activates cottonwood germination and salmon outmigration. Salmon are a major focus, but they are not the only species of concern.
4. a desire to reconfigure large-scale patterns of water use using infrastructural and regulatory approaches: metering water use at all wells and diversions, stopping illegal pumping, including connected groundwater in water rights accounting, and promoting irrigation efficiency through sprinklers or alternate stock sources.

Thinking across Imaginaries: Data, Historical Legacies, and Future Worldings

Let's briefly recapitulate Barad's theory of intra-actions in relation to water imaginaries. People's engagement with different water apparatuses (again, infrastructures, but also sources of water, climate dynamics, ecosystems, and legal and regulatory regimes) produces imaginaries of human-water relations. Water relation as a system of habits, rhetorics, policies, and norms is a Baradian phenomenon that emerges from intra-actions among entities. How people use and manage water, and for whose benefit, depends on the configuration of knowledge, law, policy, ideology, infrastructure/technology, and water—these are aspects of apparatuses, as are geophysical and ecological systems. Coconstituted apparatuses and imaginaries materialize different groundwater sciences and politics—and shifts in infrastructures, practices, politics, or scientific study—can all reconfigure the apparatus, giving rise to different flow regimes in the river and the subsurface.

An example of how two very different imaginaries can emerge from the reconfiguration of apparatuses in the same watershed is the 2008 Groundwater Study Plan, conducted by UC Davis scientists at the request of the California Regional Water Quality Control Board (hereafter Water Board). When the Scott River exceeded EPA thresholds for temperature, the Water Board identified groundwater flows as crucial to cool the stream and tasked UC Davis with determining how groundwater influences stream temperature in the valley.[56] The scientists consulted with settler stakeholders, including the county, the Regional Board, Scott River Watershed Council, and local ranchers. The scientists responded to ranchers' concerns that monitoring would lead to reduction of irrigation releases by recommending that well and streamflow data not be made public. Ranchers proposed that voluntary irrigation efficiency coupled with aquifer recharge could fix the problem, and the model was designed to test these management interventions. The 2008 Groundwater Study Plan did not recommend, or even mention, Quartz Valley and downstream tribes' preferred interventions: renegotiating water adjudications and reinvestigating the priority date of Quartz Valley Indian Community's water rights.

Public comments on the Groundwater Study Plan were heated. In protest, Quartz Valley Tribal Chairman Harold Bennett wrote, "Your draft specifically states that the North Coast Regional Water Quality Control Board envisioned the Quartz Valley Indian Community working cooperatively on the groundwater plan and we look forward to an explanation as to why this has not yet taken place. We expect you to rectify this in the future." However, throughout groundwater model development and subsequent publication of scientific and policy reports, the Quartz Valley Indian Community has never been substantively consulted by the study authors.[57] The Klamath Basin Tribal Water Quality Work Group and the Karuk and Yurok Tribes also protested their exclusion from this and other Scott Valley water decisions; the Yurok Tribe's 2018 declaration of rights for the river specifically calls out Scott Valley agricultural management. Despite these formal protests, none of the UC Davis group's publications mention Quartz Valley or other tribes or address treaty rights implications for water governance in the Scott Valley.

Commercial fishing and environmental NGOs also protested their exclusion; Klamath Riverkeeper's comments exemplify their grievances: "Siskiyou County and the Siskiyou County RCD have a well-known bias against regulation and good management of resources. Due to their anti-tribal and anti-conservation attitude many agencies have let them define studies and legal processes to get their cooperation. This is unethical and illegal as the County represents only a minority of the people [who] depend on the Scott River and often hurts rather than helps efforts such as this study."[58] In these documents and public comments, four themes characterize the distance between the two imaginaries: temporalities of groundwater recharge and salmon extinction, whether water is a public trust or a private right, whether modeling or more empirical studies are needed, and whether current practices—and not just dams, levees, and former mining—are driving salmon decline.

Under the first theme, *temporalities of groundwater recharge*, Quartz Valley Indian Community saw irrigation as the primary threat to Shackleford Creek, which flows through the reservation. Quartz Valley, downstream tribes, and environmentalists presented comments and studies showing that groundwater pumping for alfalfa had increased greatly since the 1970s, causing streams to dry earlier. In support of this conceptual model, Quartz Valley plotted monthly measurements of groundwater level from a network of reservation wells, showing that each year, spring snowmelt fills the aquifer back up, so that it sustains streamflow through early July.[59] In summer, farmers pump more and more water out, causing streams to disconnect from the underlying water table.[60] Groundwater stays cool throughout the summer, while surface water grows warmer through time. This earlier disconnection from groundwater leaves more stream reaches too hot for salmon by August and means that some streams dry completely.[61] Before the 1970s, these data show, summer streamflow was high enough, and temperatures cool enough, for spawning and rearing salmon to survive. Since 1970, streamflows and temperatures have most often been in the lethal range for salmon.

To counter this argument, ranchers affiliated with the Scott River Water Trust (hereafter Water Trust), a settler irrigator association, produced their own analyses of streamflow and groundwater data collected from members' wells. The Water Trust argued that groundwater pumping did not cause the Scott River and its tributaries to dry up during summer. The Trust's assumptions and interpretations reveal a conceptual model that is incommensurate with the salmon-centric one.[62] Key to this incommensurability is the temporal scale at which data are analyzed. The Water Trust presented well level data as annual averages rather than monthly water levels; their figures show that after the spring snowmelt, water levels reach to near the surface in most years. By definition, the aquifer is not in overdraft—a condition that describes water tables that decline for more than two years. They thus argued that, to help salmon, irrigators could reduce diversions of surface water during extremely low flows in late summer, supplement by pumping more groundwater, and be paid to do so; the Water Trust piloted this approach during the 2014 drought. Most ranchers compensated by pumping more groundwater. To justify this pumping, the Water Trust argued that pumping could continue because USGS stream gauge data showed above-average annual flows in some years.[63]

How are we to make sense of these divergent, sometimes opposite interpretations of data? Astrid Schrader's proposal for how to reconcile contradictory findings in the scientific debate over the dinoflagellate and fish toxin *Pfiesteria* can also explain the Scott groundwater case: divergences that come from uncertainties in results rather than an ontological indeterminacy may be resolved by further discussion of methods or by new data.[64] In the Scott Valley, Quartz Valley and the Water Trust based some of their analyses on the same kinds of data.[65] But, beginning from incommensurable imaginaries of water flow and irrigation apparatuses, the groups used different methods, interpreted hydrology differently, and produced vastly different policy prescriptions for recovering streamflow and salmon.

Based on monthly well level and contaminant monitoring on the reservation, Quartz Valley understands the aquifers to be highly porous, allowing flood waters to quickly percolate through porous sediments and into the empty aquifer during the fall, winter, and spring. As a result, Quartz Valley supports floodplain restoration strategies, like beaver reintroduction or mimicry, that increase groundwater recharge—but also insists that irrigation be permanently reduced, in order for those recharged groundwaters to remain in the subsurface into the summer and sustain summer flow. From this perspective, the Water Trust's approach provides no increase in benefit: ranchers get paid to release ditch water back into the stream, while still pumping the same amount of water from the aquifer that feeds that same ditch and stream.

From incommensurable imaginaries of water flow and irrigation apparatuses, the groups used different methods, interpreted hydrology differently, and produced vastly different policy prescriptions for recovering streamflow and salmon. Quartz Valley's imaginary is tied to seasonal temporalities of salmonid life histories, which require specific flows for fall or winter spawning, spring migration of coho and steelhead smolts down to the main stem and out to the estuary, and summer rearing. The Water Trust's imaginary is tied to the annual cycles of crop irrigation and multiyear definitions of groundwater policy. By presenting well levels as annual averages, the Trust's analysis smooths out the low minimum aquifer levels in fall. This data supports the Trust's contention that the aquifer is not in overdraft, because technically, overdraft means depleted over multiple years. Quartz Valley retorts that late-summer streamflow is what matters for salmon—if the stream is completely dry from early July to late September, they cannot survive.

The second theme is whether the river and groundwater are a public trust or private property right, and what standing tribes have in water and groundwater issues that affect treaty trust resources like salmon. Regarding treaty rights and tribal consultation, the UC Davis groundwater study authors and the Water Trust mentioned neither Quartz Valley nor downstream tribes' treaty rights as they bear on instream flows and groundwater. In their comments, several Native nations objected, asserting their interest in Scott Valley groundwater based on their treaty rights to salmon that spawn and rear in the Scott. The Yurok, Karuk, and Klamath Tribes representatives argued in a letter that treaty law and public trust doctrine both supported stricter groundwater regulation.[66] Commercial fishermen's organizations and environmentalists also mentioned the public trust doctrine, as a means to support coastal fishing economies and river recreation economies. The Water Trust, in its letter, strongly rejected the public trust doctrine for ground and surface water, instead arguing that regulatory action should rely exclusively on doctrines of beneficial use.

Parallel to differences over temporalities of groundwater recharge, groups held different views of the temporalities of salmon extinction. Tribes and environmentalists saw salmon's decline as urgent, predicting that salmon would go extinct within the proposed twenty- to forty-year timeframe of the groundwater study. They demanded interim results that would drive immediate management actions and called for a moratorium on groundwater pumping during the study period. In contrast, farmers and ranchers did not see their actions as proximate causes for salmon extinction; they proposed to restore physical habitat while maintaining water diversions. They wanted to wait for long-term study results before restricting groundwater pumping. They agreed to participate in studies only if researchers assured them that the model results would not be used to reduce irrigation deliveries or mandate conservation measures. Tribes saw this as a delaying game and demanded regulatory action based on existing streamflow and groundwater data.

The third theme is what kind of science is needed to understand groundwater–surface water interactions and to govern groundwater and surface water as connected units. Ranchers and UC Davis modelers asserted that large-scale models approximate aquifer dynamics well enough to be used to propose groundwater management actions.[67] These actions are generally focused on enhancing recharge or efficiency, based on the idea that groundwater use does not reduce river flow.[68] Ranchers and their elected representatives eschewed empirical analyses, water metering, and mandated well monitoring. Study authors did want more data to improve their model, but they rejected the idea that well data should be publicly available.[69] The different tribes and environmental groups all advocated for analyses of empirical data, as adjuncts to models. They wanted to know exactly which wells are connected to the stream so pumping from those wells could decrease while modeling continued. Quartz Valley proposed—and later conducted—a simpler analysis of historical streamflow data from USGS gauges for changes in days with low flow, correlating those flow changes with groundwater pumping over the last half century.

The fourth theme is the significance of so-called legacy impacts, like dredging and levee building. Ranchers and county officials see these impacts as primary drivers of salmon decline and see restoration to baseline conditions as unrealistic. Instead, they propose beginning from a baseline of current conditions and working for incremental improvements in habitat conditions where landowners see a benefit. Quartz Valley and the coalition of downstream tribes and environmentalists see legacy impacts as significant and also ongoing—consistent with Indigenous analyses of colonialism as a continuing process.[70] They argued that to prevent imminent extinction of several salmon runs, restoration should maintain enough cool water in the stream during critical periods when large numbers of salmon perish. For these groups, putting water back in the stream is far simpler than reengineering large areas of floodplain.[71]

In comparison to the top-down management policies that I sketched earlier in the chapter, both beneficial use and public trust imaginaries in the Scott Valley are metaphorical underflows that challenge dominant agency science and legal discourse about groundwater, streamflow, and salmon. The Regional Water Board's TMDL Action Plan produced substantial support for the idea of building a scientifically robust model to shape policy, but actors disagreed on nearly on all aspects of modeling. But as these groups were debating restoration approaches, new science about beavers' ability to recharge groundwater and create coho salmon habitat reshaped these debates and forged new alliances in the Scott.

Top-Down Processes and Underflows—
How Imaginaries Shift in Places

As fish- and farm-centric imaginaries have converged around groundwater recharge via floodplain reconnection, once-unimaginable partnerships seem possible. In 2008, during the groundwater study planning process, the Karuk Tribe was among those criticizing the Siskiyou Resource Conservation District (RCD) for its anti-Native bias. A decade later, Karuk Fisheries and the RCD worked together to build off-channel habitat ponds on a lower Scott River ranch. All parties agree that this stretch of river is crucial to recovering coho and Chinook salmon. Recognizing Karuk expertise in constructing off-channel ponds, Siskiyou RCD brokered the relationship with the landowner, then subcontracted out the project. In 2015, Quartz Valley and the Watershed Council had begun collaborating on a summer program that brings tribal and nontribal youths together to do watershed restoration work. The SRWC–Quartz Valley partnership grows from their mutual interest in recovering salmon while developing local economies and infrastructures so that they are resilient in the face of climate change and decreasing snowpack. One topic of shared interest is beavers' improvements to salmon habitats, and shared interest in understanding how much beaver dams (and human-made "beaver dam analogues") can recharge aquifers. They now share data and are collaborating on funding proposals for stream restoration. That such collaborations are happening at all represents a marked shift in Scott Valley politics. But whether these temporary alliances will shift underlying imaginaries remains to be seen. A 2018 SRWC Planning Report attributes stream drying to legacy watershed degradation, drought, and climate change, not groundwater pumping or surface diversion. The SRWC report does not mention the Quartz Valley's core issue of regulating, metering, or readjudicating surface water diversions from tributaries. Quartz Valley shares the SRWC's goal of improving habitat and recharge using beaver dam analogues and other techniques—while also taking policy and regulatory action to get more water into the stream directly.

Regulatory changes have accelerated since the 2011–15 drought. The 2014 State Groundwater Management Act for the first time mandated that groundwater be monitored and managed by a group comprising all interested parties. For the first time, Quartz Valley has a seat at the table in settler governance of water in its homeland. The drought created new funding and regulatory mechanisms, facilitating floodplain restoration projects in the Scott. The drought also spurred experimentation, in flooding fields in spring to recharge groundwater and in water banking. The Scott River Water Trust, a rancher organization, set up a financial structure to lease water from ranchers to increase summer streamflow. Their rationale was that ranchers would use the payments to pump groundwater instead. This project relied on the idea that this groundwater was not connected to streamflow—a facet of the hyporheic imaginary with beneficial use discussed earlier. Ranchers have long held this imaginary and describe how the groundwater aquifer refills each spring as snow melts off the mountains, then empties during the summer as they pump irrigation water; the UC Davis groundwater model validated this conception, finding that the aquifer was not technically overdrafted.[72] (Statutory law defines groundwater overdraft as a lowered water table that persists for more than two years.)[73] In the water trust proposal, farmers leave surface flows (which would normally flow through their ditches to irrigate alfalfa fields) in the stream, receive payments from the Water Trust (using state funds), and then use those funds to pump groundwater to irrigate their crops.

Quartz Valley, concerned about streamflows in Shackleford Creek, presented an argument that rests on an imaginary that challenges settler governance practices of managing groundwater, streamflow, and salmon separately: the hyporheic imaginary with multispecies commons. The river dries up each summer because ditch water diverts most flow and groundwater pumping sucks up the rest. To support this analysis, Quartz Valley drew on similar data to the Water Trust, but interpreted it differently. Although water tables refill to the surface each year (the Water Trust's argument against overdraft), the summer water table has been dropping lower and lower over the last half century. To sustain salmon and their multispecies commons, this hyporheic imaginary requires that ground and surface water be managed together and considered as a tribal trust. Quartz Valley has exercised tribal sovereignty to secure more water by petitioning for "treatment as a state" under the Clean Water Act, which gives the Tribe the power to set temperature and streamflow standards. It is the latest strategy in Quartz Valley's struggle to manage salmon throughout the Scott Valley, by demanding changes in irrigation and enforcement of state fish and wildlife codes. Until 2014, Quartz Valley was shut out of county-run groundwater governance in the Scott Valley, until the California Sustainable Groundwater Management Act mandated all parties' inclusion in the new groundwater sustainability agency.

This sketch of Quartz Valley's multifaceted environmental protection work illustrates the circumstances in which many Native nations work, both those that are federally recognized and those that are not. They make distinct but allied demands to settler governments to return stolen land and comanage rivers, landscapes, and fisheries, while dedicating resources to science that grows from and supports cultural relationships. Thinking with Kim TallBear, Kyle Powys Whyte proposes that Native nations' ecocultural resurgence projects "are invested in more expansive objectives of indigeneity that seek to undermine structures of industrial settler colonialism. Indigenous or other persons who are not members of the particular Tribes mentioned, or members of settler populations, can and are expected to learn from and contribute to these projects though they do not, of course, somehow become members of the particular Tribes by doing so."[74]

In California and beyond, recognized and unrecognized tribes are expanding the lands that they manage, by buying land outright, pushing cities and state agencies to return land, forming land trusts, working with nonprofits to manage community forests and prairies, and negotiating cultural use areas within state parks.[75] Tribal treaty rights guarantee comanagement and the right to access and manage cultural and natural resources in tribal territories. Where reservations were established after the treaty era, as with many rancherias in California, federal and state law and policy support tribal sovereignty and government-to-government relationships. We settlers must push our governments to uphold these commitments to Indigenous sovereignty and must work directly to support Indigenous ecocultural revitalization, if incipient alliances are to produce new ways of living together along rivers.

Coulthard and Simpson's framework of grounded normativity argues that settlers in Native alliances must follow protocols and legal orders that flow from "[Native] relationship to the land itself [that] generates the processes, practices, and knowledges that inform our political systems, and through which we practice solidarity."[76] Placing Whyte's formulation of indigeneity as ongoing work to undermine settler transfiguration of landscapes alongside Coulthard and Simpson's formulation of grounded normativity provides an orientation for doing this work.

Some alliances between settlers and the Native nations on whose land they live have started to transfigure those surface flows from below. But most non-Native people don't consult local Native nations as they manage land, water, wildlife, and other natural and cultural resources; even fewer return land or engage in formal comanagement. This settler colonialism as ongoing dispossession is evident in Salmon Creek, where residents and local scientists (mostly white) erase Pomo and Miwok sovereignty by not formally consulting with the Federated Indians of Graton Rancheria.[77] When asked why, one resident responded that Graton tribal members had lost their culture; another said that because tribal members who revived cultural practices consulted anthropological records, they were not authentically Indigenous. J. Kēhaulani Kauanui's theorization of commonplace, everyday erasure is useful in understanding this dynamic in Salmon Creek and across California, where settlers mobilize racist purity logics that see "living Indians as 'mixed' and therefore no longer 'truly Indian.'"[78] This insistence that Indians have vanished despite their continued presence is especially acute in California, where Congress either declined to ratify many treaties or later dispossessed tribes of land and recognition.

How to Practice Unsettling in River Restoration

I first came to the Klamath in 2004, to interview Klamath Tribes fisheries staff and upper basin ranchers in the wake of a 2002 fish kill that precipitated the current dam removal effort. I found points of convergence around conservation ranching, Native forest management practices, and concern for salmon.[79] In 2011, my classmate Daniel Sarna-Wojcicki and I found a similar convergence in the Scott Valley around the idea of beaver-led restoration, and we returned over the years to interview ranchers and Watershed Council members. In 2017, now a professor at UW Seattle, I began formally collaborating with the Watershed Council and the Quartz Valley Indian Community through a field course that brought together UW students with Karuk and Quartz Valley high school interns to survey salmon, measure water quality, and explore groundwater-stream interactions. In the Scott Valley, my students and I are currently working with a team of economists, ecologists, and hydrologists to develop a computer model of groundwater pumping grounded in distinct imaginaries. With this model, we can explore the long-term stability of farming profits and fishery growth and propose groundwater pumping schedules that leave some water for salmon. By first conducting interviews and observing Groundwater Advisory Committee meetings, we are cogenerating specific model architectures and inputs guided by Quartz Valley, the Watershed Council, and a rancher group's priorities; these interviews are revealing nuances to salmon- and cow-centric imaginaries.

Without cross-cutting partnerships, rural landscapes are ill-equipped to face devastating flood, drought, and wildfire. With such partnership, tribes and non-Native managers can work to innovate by revitalizing cultural practices to reimagine and reinvent cutting-edge climate adaptation tools. Crucially, these partnerships must create relations and conditions of trust to break through stalemates of fragmented governance. Those processes with the highest level of community support remix Western science, Indigenous science and knowledge, and/or local knowledge, because when people shape research questions and participate in analysis and interpretation, the results are more relevant to their problems and concerns. In the language of feminist science studies, these different ways of knowing are "situated knowledges"—each embedded in a cultural context; bringing these knowledges together can contribute to a stronger understanding of a given problem than using Western science alone.

Afterword: Beavers Remix Water and Its Role in Sustaining Riverine Species

As restoration practitioners and "beaver believers" work with beavers, their encounters sometimes reshape imaginaries. Beavers remix spatial distributions and temporalities of wet and dry, blurring the boundary between land/water, channel/floodplain, surface/groundwater, now/later, and human/nonhuman. In so doing, beavers can inspire new modes of water governance and new ways for humans to relate and belong to watershed communities. At times, close observation of beavers leads people to reconceptualize stored and infiltrated groundwater as water that supports habitat for others now and emerges downstream as habitat for others later in the season.

In the Scott, some of these new imaginaries appeared in the 2010s, after beavers reconfigured water flow with the help of human collaborators. A rancher didn't have time to drive his bulldozer down to the creek, as he did each summer, to tear out a dam that beavers had built. That fall, he noticed that the creek downstream of the dam was flowing for the first time in decades. He alerted the Watershed Council, who were intrigued by beaver-assisted restoration after hearing a talk by ecologist Michael Pollock. They invited Pollock to visit in 2012, to present data from his beaver experiments in eastern Oregon. He spoke to a packed house of ranchers and tribal members, who all seemed excited about the benefits of more beavers. Public comments from a Siskiyou County hearing a few months later show that beavers had entered the hyporheic imaginaries of ranchers and salmon advocates alike. Marilyn Seward, then chair of the Watershed Council and Etna City Council member, advocated for increasing beaver dams as a groundwater recharge strategy by referencing Beaver Valley. "Scott Valley owes their wonderful soil and everything else to beaver. It was Beaver Valley before it was the Scott Valley," she said, adding that many landowners supported working with beavers.[80] This description evokes a more-than-human commons created by beavers before white settlement—a lush world that could potentially be re-created by settler descendants if hundreds or thousands of beaver dams collectively raised the water table beneath the valley.

Yet not everyone shared these sentiments. At the same meeting, a trapper for the US Department of Agriculture voiced concerns about the potential negative impacts of beaver dams due to bank erosion and the possibility that "beaver populations [would] get too big and need to be thinned."[81] These discussions about the moral and legal status of beavers were directed through the prism of human interest. For example, one county supervisor remarked that he was "skeptical of the effectiveness of beavers as a watershed tool" because he had "heard negative stories from several people about the animals interfering with irrigation ditches and other agricultural operations."[82] All of these comments instrumentalize beavers, whether as useful tools or harmful disruptors. "Several landowners that were present at Pollock's presentation are interested in exploring the possibility of encouraging beaver that are already present in the Scott River to build dams in strategic locations where it could primarily benefit the groundwater [and irrigation practices] but also the riparian vegetation and fisheries habitat," said Watershed Council project coordinator Danielle Yokel, further illustrating this point.[83]

Pollock and the Watershed Council built three experimental beaver dam analogues to study how they affected the stream and its young salmon. The experiment spawned the first collaboration between the Watershed Council and the Karuk Tribe; the council invited the Karuk fisheries department to tag fishes at the Sugar Creek beaver dam analogues. Before this, the two groups used different restoration techniques. Karuk Fisheries dug off-channel ponds as salmon refuges, while the Scott River Watershed Council built beaver dam analogues. Through conversations around the fish-tagging table and field trips to one another's projects, both have expanded their repertoire of restoration techniques. By sharing methods and data, they have developed a more robust understanding of how groundwater influences stream temperature, which in turn shapes fish behavior.

In the ensuing years, the Scott River Watershed Council has worked with ranchers throughout the valley to trick beavers into moving their dams away from irrigation diversions, to protect trees from beaver chewing, and to partly drain beaver ponds that flood fields or infrastructure, as well as to install beaver dam analogues that often are maintained by local beavers.[84] In an educational video aimed at her fellow ranchers, Betsy Stapleton, now board chair, describes managing beaver in a similar way to managing cattle, deer, or elk—planting willows for them to eat, directing their activities to specific areas of the ranch, and decorating their house with their artifacts (chewed logs in addition to racks of antlers).[85] Such narratives seek to engage ranchers in salmon recovery by presenting beaver dams as a means to maintain a settler-dominated landscape.

As more ranchers switch from trapping beavers to building dams with them, a beaver-inspired hyporheic imaginary may transfigure knowledge making, water governance, and earth repair.

Underflow 2

Being in the Field

I had a body before gender. My perception of my body emerged co-constituted with LA's dry culvert streams, with back lots, and with small strips of waste grass along the fences of asphalted school yards. These were spaces where, alone, I had no gender but just a body, not seen in a mirror, but rather experienced through touch and contact with dirt, grass, and water.

In later fieldwork, I added a choreography, a rhythm, to this foundational method of observation. In Montana, studying geology as an undergraduate, I learned the random walk method of pebble counts, bending, immersing neoprene hand in ice, touching a pebble behind my right foot without looking, measuring its intermediate axis, calling out the measurement, replacing the rock, taking a step in a random direction, bending again. This choreography embedded a relation of noticing into my muscles and skin. Studying how the reservoir sediments washed downstream after the Milltown Dam removal on Montana's Clark Fork, I first practiced movement as measurement, freezing as I dipped my hand into the water to grab a stone and leaned into oar-strokes in an inflatable raft on the Blackfoot arm of the Milltown reservoir. I rowed while another student ran an echo sounder to chart the bottom. This was in March, when slush formed on white-capped foam as the wind pushed us sideways; later, we had to correct the profile of the reservoir sediments to account for our sinuous path.

My thrill at these bodily encounters with rock and water was tempered by social anxiety. On big rivers like the Clark Fork, geomorphology field campaigns require teams of people, and all of the social interactions around these field days were intensely gendered. I was not out as trans and so got read as cisgender male, but off. My self-identification was more genderqueer butch, but where does one bring these nuances up amid classroom discussions of crystal structures and physics of groundwater flow models? In my lab, there were no women, no other trans people, no other queers. Field and office conversation tended to heterosexual locker room banter (rating female colleagues' sex appeal, describing sexual and drinking exploits, and the like). While programs for gender equity in STEM at least made some space for solidarity among women-identified scientists, I was afraid that being out as trans would expose me to exclusion or violence or would close off access to scientific opportunity, so I opted to pass and take the cuts and scars invisibility inflicts.

I escaped to the field, more often than not, solo, where gender as a social category receded some. The feeling was like peeling my wetsuit off after a day of snorkeling to count fishes: a lightness, a lack of constriction, a dissolution of a sharp boundary between my body and the world. Solo, I sidestepped the social cues that signal right and wrong gender. I evaded rejection and threats of violence and focused on counting cutthroat trout in a backwater pool that formed in the reservoir sediments after the dam came down, amid the warm summer wind sweeping down the Blackfoot River canyon. At times, in the field with others, the focus and choreography of observing sediment deposition and scour or wading in icy currents with a surveying rod meant for a while that we were all just muscle and skin, adrenaline, pain, and sensory overload.

# 2

Queer × Trans × Feminist Ecology

Toward a Field Praxis

> The insight we draw precisely from decolonial feminisms, Indigenous studies, and trans of color theory is to understand "theory" differently: not as knowledge that issues from within the academy or that aspires to academic recognition but that invents itself on the fly, in the midst of a campaign, in the telling of stories.
>
> —AREN AIZURA ET AL., "Introduction: Decolonizing the Transgender Imaginary"

> Describing the depressive position in relation to what I am calling "brown feeling" chronicles a certain ethics of the self that is utilized and deployed by people of color and other minoritarian subjects who don't feel quite right within the protocols of normative affect and comportment.
>
> —JOSÉ ESTEBAN MUÑOZ, "Feeling Brown, Feeling Down: Latina Affect, the Performativity of Race, and the Depressive Position"

Introduction: Study Site: What Body Studies?

Field time is an open-ended time. Scientists switch modes when rain, spawning, bloom, or a drop in temperature signals it's time to head to the field—same as when cruising for sex, you'll miss out if you don't know where or when to be. En route, in cars, in boats, or on trails, we speculate about broader questions and patterns of research much more wildly than in any formal talk or paper. We talk politics, talk shop, or reminisce as we clamber down precipitous, bramble-covered slopes. But once we reach the site, such talk stops. Someone calls out numbers, others discuss instrument calibration. We pause to describe an organism's identifying characteristics to students. This methodical attention makes for precise, even painstaking measurements. At times, we fall into a rhythm, each to one's allotted task, moving together with practiced, almost rote movement, yet this rhythm is not a factory's strict clock time. In the field there is time for distraction: a fish biologist admires a snake crossing the road, an entomologist tracks birdsong. Once, a botanist friend helping record fish count filled the margins of my data sheet with Latin names of plants. Our love for the field's multiplicity exceeds quantification. In its repetition and excess, field time is like cruising time. In the metaphorical river of underflows, we are beneath the surface of the gravel, cruising the hyporheic zone while doing the part of science that is not seen in published papers but wells up into the surface in conference talks in joke slides or self-portraits.

Time in the field demands bodily effort, and it often damages our muscles, joints, and tendons. As a field scientist, I catch fish in two-pole seines, bent over, arms extended in a pose that leaves my arms, shoulders, and back in spasm. I drive pipe and rebar into gravel bars with a rock or sledgehammer until my hands and arms are numb. My students and I crawl up the stream for miles, in wetsuits and snorkels and masks, necks craned to spot small fish in the shallow streams. In the field helping labmates studying trees or soils, we bend down to pick up a rock or a leaf, kneel in gravel or duff to collect bugs or sand or seeds or identify flowers. We chew on rocks and green plants and taste waters for hints of salt or acid. We do this over and over again, for days. Sometimes, we carry backpacks of instruments far into the backcountry. Over years of repeated exposure, these labors cut and scratch us, wear out our joints and leave our skin weathered and cancerous.

In "Six Ways of Looking at Crip Time," Ellen Samuels writes about the bodymind's resistance to becoming "someone . . . whose inner clock was attuned to my own physical state rather than the external routines of a society ordered around bodies that were not like mine." For me, and for many who continue this work for decades, field time becomes crip time—fieldwork slows once one can no longer perform its movements because of pain, crushed disks, or a third set of knees.[1] Instead, we rely on assistants, who seine fish, crawl through brush and up streams, and stoop over and over to sample plants, soil, rocks, or water, but inexpertly, the movements not yet imprinted into their muscles. The field changes then, following a pattern that Samuels describes: "For *crip time is broken time*. It requires us to break in our bodies and minds to new rhythms, new patterns of thinking and feeling and moving through the world."[2]

Disabled people share with queers of color the experience of being "cast out of straight time's rhythm" Muñoz's response to Tavia Nyong'o's question, "Is there something [B]lack about waiting?"[3] In this circuit of waiting and feeling, I explore how lived experiences of difference make different attunements to the field—and how alternate field temporalities in river sciences make new questions and provocations.

I trace fishy feelings from the field to sites of scientific performance (the conference, the journal article, the policy debate), laying the groundwork for later chapters' deep dives. To explore these sites' queer temporalities and affects, I draw on my experience studying rivers in the western US. I describe a "particular affective circuit" that springs from the "brown feeling" that Muñoz theorizes in "Feeling Brown, Feeling Down: Latina Affect and the Depressive Position."[4] I want to extend this "depression that is not one" to think about affects that pass among humans and nonhuman beings we meet in the field as transspecies melancholia, a trans way of making reciprocal relations of care with the more-than-human world. To tap into these affective circuits requires that scientists take on a depressive position that might be a fishy, creaturely, mushroomy, microbey, or glaciery feeling.[5] They might do so through public, affective engagements with species facing extinction. Such practices and positions resist normative affects—the white, male, straight, cisgender ways of communicating feeling that dominate scientific institutions and forums.

Being trans or queer in the field—like being Black, Brown, Indigenous, or female—carries risks of violence and exclusion. Such positionalities can lead to different choices for field sites, human collaborators, and ways of relating with organisms that disrupt colonial practices of field science (though they by no means always do so). Black people, Indigenous people, and people of color in the environmental sciences have pushed for new research agendas that focus on racialized communities' environmental engagements, on urban environments, and on racist policies.[6] They often foreground queer and feminist tactics to do this work, sometimes making

Figure 2.1. Susannah Maher (left) and Kimberly Yazzie (right) record salmon habitat data. Charna Gilmore (center) evaluates how much cover streamside vegetation provides to young fish. Photo by Michael Bogan.

alliances with white and settler scientists to create supportive spaces, within our labs and professional societies. Solidarities forged through these alliances have developed antiracist and decolonizing methods and spawned scientific collaborations. Their most crucial function, however, is to make scientific work survivable for we who feel otherwise.[7]

To explore these potentials, I use field writing tactics and Gloria Anzaldúa's strategy of writing "autohistorias"—personal reflections on the experience of being an outsider or border-dweller that crystallize the political stakes of challenging racist and heterosexist institutions.[8] Inspired by Muñoz's New York City field walks in "A Jeté Out the Window,"[9] I visit sites of ephemeral presence and enact performances as methods for tracing affective underflows to dominant scientific discourses in field scientists' expressions of multispecies feeling.

Eve Kosofsky Sedgwick wrote that endurance to write a particular thought comes out of one's *different* body, shaped through its movements and relations, always influenced, but never fully determined, by race, gender, class, and sexuality.[10] Sedgwick hoped that other writers would trace their embodied commitments thus: What thoughts, books, and/or projects reach out and grab one's attention? How do those ideas resonate with confrontations in the streets, relations with dying loved ones, encounters with state power and sanctioned discourse? As a field scientist, part of my response to Sedgwick's call is to tell stories of how conceptual underflows—this time, field temporalities—seep up into conference presentations and journal articles. Through autohistorias, I trace my urge toward a trans-queer-feminist ecology through a series of encounters, in the field, in conference rooms, and on social media during the fall of 2014.

Field Methods: Autohistoria: The Ecology Conference

In a crowded meeting room, windowless and stuffy, a woman explains a numerical model of yellow warbler migration. The model looks like a map of North and Central America, with colored dots of different sizes representing the bird's summer and winter habitat. She presents mathematical methods and results of different models. The yellow and red dots grow and shrink, in response to different scenarios of habitat protection and destruction in Central America and the southeastern US.

The bird itself does not appear, neither in likeness nor in song. She does not describe the early dawn counts, the flash of yellow through mist, or birdsong piercing the insect hum. Perhaps to the other field scientists in the room, with their own intimate encounters with their study species behind each of thousands of data points, those intimacies are present tacitly. The model results, she says, demonstrate that unless all of the remaining Central American forest remains standing, the warbler will go extinct. But the end of logging, she says, is unrealistic. There is some tenseness in her shoulders as she says this, and a tightening of the muscles around her eyes. My chest goes tight, and I am thinking all at once of the sound of chainsaws in a Guatemalan forest, a young guide under branches cloaked in monarchs in a Mexican butterfly sanctuary describing how comparatively few now returned, and small fish gasping for oxygen in shrinking pools near the California coast. I am thinking of how Judith Butler writes about grief, and grievability, and what things or people we are allowed to grieve for—those questions feel like pressure behind my eyes.[11] Are all of the scientists in the room—seemingly bored, or tired, or engaged in the data—actually mourning their own loved, near-extinct others, whose deaths are not grievable here?

In straight ecology, field experience is rarely allowed to transgress into the conference, and affect is regarded as wrong or queer. Was the yellow warbler scientist thinking about her field site? I was thinking about my field site. But if the field comes to other scientist's minds as they give their conference talks or write their papers, it does so tacitly. I too often do this—I privilege data over feeling. But data are imbued with feeling, and that feeling can change data's meaning and how it is taken up in river management. If people's relationships to land and other species were recognized in the field as emotionally significant and reciprocal with their study ecosystems, those fifteen-minute talks could discharge some of that grief and could forge connection. This is my dream for a Muñozian, utopian ecological science, which, like queer, is not yet here.[12]

If the field were queer-trans-feminist, then affects that flowed between me and my study species could go into my own presentation at that conference: Here is the cold fish, anesthetized, not flopping, gently between my fingers; here is its slime mixing with creek water on my hands; here is its weight, fifty grams, its small life spark there. Here is the little fish hiding under the branches; here is my back aching and my shoulders cramping. Here is the gravel digging into my knees; here is my grimace of concentration. I balance all these desires—my desire to collect good data, the fish's desire to flop away, my fish-catching mentor Michael Fawcett's desire for coho and steelhead to keep returning to the creek in front of his house and for ethical fish sampling practices to become the norm. Kneeling in Fay Creek, the hard stones and cold pushing through a hole in my waders that barbed wire made as I crossed onto my field site in a cow pasture on private property—all this is here in my scientist body, but it is not something I feel I can speak of as a graduate student in my own allotted fifteen-minute slot at the conference.

I don't fit in at the conference, not only because I am queer and trans but also because I need the field to be here in the conference room. I imagine that many straight, cisgender scientists also wish the field could come in, but they seem to bear its absence better. Most of us became ecologists because we love the field. I chose my case study as a way to get into the field in relation to people who lived there and thought about the field. I recognized early on that without these field relationships to sustain me, I would not survive the academy or grappling with salmon death.

Field Methods: Autohistoria: My Dead Cutie, 2014

In late August, my lab-mate Mike Bogan, an aquatic entomologist and well-networked Twitter user, is rooting for team benthos in a #cuteoff. The rules: original photos only, and you must say why your study organism is the cutest. I post some stills of baby salmon that I took in June, while swimming through a deep blue pool teeming with coho salmon and steelhead trout fry. Over the next weeks the photos flood in. Most of the animals are small or big-eyed. Many of the small ones (bugs, frogs, snails) perch on a person's finger or nestle in their hand. The scientists who post also speculate: What makes them so cute? Their small vulnerability? Their big infantile eyes? Other photos, of sharks, elephants, and slime molds, are neither small nor traditionally "cute," but their researchers argue for their inclusion nonetheless. These images, of a researcher and an elephant entwined in the water or a smiling scientist tightly hugging a toothy baby shark, transpose traditional gestures of human intimacy onto strange intimates. Many nonscientists who comment on the posts seem surprised by this outpouring of love from scientists, even within the limits of "cuteness"—a queer aesthetic when applied to slime mold—and Twitter's 140-character limit. Scientists call the #cuteoff a victory of science communication. But I wonder, what is being communicated through these images? What is being said, how is it being heard, and what affects are passing from person to person as the photos and characters fly across the ether?

The trope of cuteness seems a deliberate rejection of doom-and-gloom environmentalism, which has, in recent years, fallen out of favor as a way to convince the public to conserve biodiversity or fight climate change. Yet in excising doomsday narratives, a space for mourning the "double death" of human-driven extinctions is also foreclosed.[13] Or perhaps not: the response by nonscientists suggests a shared recognition, in cuteness, of something worth looking at and worth posting about. I read this response as a kind of anticipatory nostalgia for cute, biodiverse field worlds so far from the Twitter users' screen-bound time: a nostalgia for worlds that are not yet quite gone.[14]

The next month, my lab-mate Suzanne joins me to count salmon fry on Salmon Creek. The deep pools I swam through in June have shrunk to reddish puddles separated by long expanses of dry streambed. Suzanne and I stand there on a gravel bar above a small pond of reddish water; she measures the dissolved oxygen with a handheld probe. The palette is all gold and gray, storm clouds, alder leaves, and then this strange red-purple that turns the water opaque and sulfurous. I switch on a flashlight, crawl toward the edge of the water, and then slither into water barely belly deep. I scan through the murk for a flash of silver. I see just one slowly moving fish through the dim.

As we survey pool after tiny pool scattered along the dry creek bed, cold seeps through the shredded knees of my wetsuit and around my snorkel, bringing decay onto my tongue. I strain for any glimpse of fish. It's not just the low light or the way the redness attenuates my flashlight beam: my heart aches for the hundreds of darting silver bodies I swam through in early summer, for the lives gone missing. In one pool, nine small silver fish flit past. My heart leaps—but then the red pulse of their slow-gasping gills sets up a resonant feeling of suffocation as I struggle to breathe through the snorkel. Here Suzanne measures dissolved oxygen at four parts per million, well below the six parts per million salmon need to thrive. In the next pool, strange purple biofilms bloom up toward the surface; here the oxygen is close to zero, and we find only sticklebacks.

I dive again and again into putrid water as Suzanne measures oxygen. In one pool, strange purple biofilms bloom up toward the surface; here the oxygen is always close to zero, and we find only three-spined sticklebacks. At another pool, the water is opaque and black with cyanobacteria. Suzanne kneels down and picks up a small dead steelhead in her neoprene-gloved hand. "There must be a lot of morts," she says, using jargon for *mortalities*, or dead fish. "You usually don't see them before the raccoons get to them." The little body is fresh and stiff, its spotted side covered with a whitish film of microbes, its bright eyes glazed over. "So small and skinny," Suzanne says, "compared to the streams we survey [further north, in an ecological preserve along the Eel River]. They're just losing weight in these small pools." Because of the drought, the stream stopped flowing a month earlier than usual, and it's been months since the fish's bug prey drifted downstream. In responding to this single salmon death, Suzanne seems curious and slightly wistful, as if she is seeing the liveliness stilled, a kind of sadness at the lost possibility of smolting, ocean swimming, and a chance at spawning. But alongside that sadness at the individual fish's death, we both wonder how other species responded to that small body, how nutrient cycles might shift in that pool with the addition of its carbon and nitrogen. We are fascinated with the relations these deaths and drought revealed. In the field that day, we talk about that little salmon's role as ecological agent, as raccoon food, and as a sentinel of changing hydrologic conditions. We wonder whether the young fish that survived had sought out the sanctuaries that stayed wet and oxygenated through the long and extraordinarily dry summer. If so, how? Did they smell the groundwater flow? Choose the deepest pools? Or did they end up there because other fish drove them out of the other pools? Mixed with grief, I feel and sense that she feels some joy in that messy life and death that drives adaptation and selection. This drive to flash and shimmer, to flick a fin and gulp a bug, to dart in a current and swirl away from a shadow to a hideout under a bank—these little fish are cute, sure, but they're fierce, too, and they remind us why we fight for them with defiant, glittering gestures.

I pick up one dead silver body and photograph it, thinking to post it to the Twitter #cuteoff. The silver fish body in black-gloved hand reproduces the #cuteoff setup, only instead of a circuit of liveness between bare hand and live creature, neoprene stretches that distance. The dead fish and I touch only through the water that seeps in around my wrist. In my bare hand, it is still and not slimy, and it weighs almost nothing. I know these young fish partly through their jumpiness, their determined flopping when I grab them from a bucket or net to weigh and measure them, the pulse of their terror even when confined in my wet hand. This lively cuteness is bound up with grief and mourning; the youthful exuberance is all the more poignant because it is a life cut short. But saying this publicly, especially with the hashtag #mydeadcutie, seems a transgression of something unspoken about loss, or a breaching of the particular affective circuit of fish and human in the field. I wish I could cry like Nao Bustamante in *Neapolitan*. Instead, fearing trolls and heartache, I don't post the photos.

Theoretical Approaches: Queer Time, Field Time

> So, what's queer about waiting? There is certainly something familiar about it to queers, whether it is waiting for sex, love, reciprocity, freedom, justice, or any of the goods without which ordinary existence feels that much more precarious. The degree of willingness to wait is often the measure of the difference between pedagogic time on the one hand, deferred and deferential, and

performative time, insistent, repetitive, and filled with "the presence of the now," on the other. And yet, the relation of the queer subject to waiting cannot cleanly be divided between refusal and acquiescence. Part of what's queer about waiting, the part queers seem to recognize in it, lies in its imaginative and even generative contours. Anticipation, for example, seems a fruitfully queer position, given its potential deviations from straight time. This suggests that waiting can draw out an oscillating tension between the probable and the possible, disclosing behind its apparent inertia a rhythm and, even, a music. Waiting can insert a pause between past and future, a pause in which the subject's relation to both can be, as it were, performed.

—TAVIA NYONG'O, "Brown Punk: Kalup Linzy's Musical Anticipations"

In response to Tavia Nyong'o's questions, "What's queer about waiting" and "Is there something [B]lack about waiting?" Muñoz replies "Yes!" and contends that waiting also links queer, Latinx, transgender, disabled, Indigenous, Asian, and poor people's experiences of time: "We have been cast out of straight time's rhythm, and we have made worlds in our temporal and spatial configurations."[15] Field scientists, overwhelmingly white, straight, and not poor, have mostly not been cast out of straight time in the same way as Muñoz proposes; Kalup Linzy, a Black queer drag performer who sings from his couch about waiting for a piece of trade, performs, in Nyong'o's analysis, "a radically passive refusal of the affective limits of homonormativity, one that reconnects the subject to more vagrant and disreputable resources of queer history."[16] In asserting that such vagrant and disreputable resources might be the medicine that ecologists need, I begin to outline a field praxis that both resists straight time's strictures and makes space for more Black queer scientists in the field. I make this perhaps jarring leap from queer performance to scientists' field sites by means of an observation: in reflecting on their field encounters with threatened and dying species, field scientists enact some of the qualities Muñoz finds in Linzy's and his contemporaries' queer performance: "practiced failure and virtuosic." Scientists' field movements and conference performances are virtuosic in detail, mathematical elegance, timing, and argument. Unlike Linzy, these scientists don't deliberately perform failure, yet in private we ecologists often lament our collective failure to avert extinction and climate change. Like living a closeted life, does this private lament cause psychic harm?

I wonder if this hopelessness felt by ecologists could be transfigured into a Muñozian "hopeless hopefulness" by thinking queerly about histories and futurities of ecological science.[17] Such an effort would entail "a politics oriented towards means and not ends."[18] Within capitalist logics of growth and resource substitution, there is little time to attune to different temporalities—species' life histories, carbon cycles, geologic time scales of warming and cooling, uplift and erosion. And yet these attunements, in my own queer ecological field practice, disrupt predictions of salmon extinction with the lively emergence, each April, of tiny big-eyed salmon fry gulping bugs with utopian exuberance. This darting liveliness is an interspecies affect that travels to me from the little fish. In so doing, it reshapes my field time as an escape from neoliberal inevitabilities, a rebellion against the increasing parceling out of time that limits daydreaming, open-ended speculation, and intellectual surprises. Inspired by tribal and environmental justice scientists' innovations to center science on community priorities for repairing environmental harm, I explore whether science grounded in such queer-trans insights and temporalities can make a different move in a similar direction, a turning along a shared path.

Like queer time, field time is characterized by repetition: as we revisit the same tree or field or stream season after season, we perform mundane motions of care for data and species methodically, with practiced motions. As we measure temperature, weather, erosion, or the condition of plants and animals, we are searching for a combination of characteristics that indicate healing or decline, as human caretakers do for sick and dying loved ones. That these measurements are expressed formally as statistics in mathematical models does not take away from the feelings we have for these companions. The end here is representation in statistics and models. But the means—field practice—offers an opening to change the conditions of possibility for species and ecologies, by expressing feelings for nonhuman others, thereby deepening our relations with them and with one another.

Like the mundane movements of farm work, factory work, or housework, the repetitive motions of fieldwork become second nature and unremarkable, leaving us free to notice a passing butterfly, a bloom of strange microbes, a stream-side apple tree heavy with fruit. Like scribes of old taking dictation on paper, or information workers who input data into devices, we become adept cyborgs, attuned to gaps, typos, and mistranscriptions. Within the neoliberal university, which limits social time and delegates field time to students, field time is becoming queer in that it has been cast out of straight time's rhythms. Refusal of straight time's rhythms is already strongly articulated in ecological science by two kinds of scientists who, before becoming scientists, were already (perhaps) cast out: Indigenous fish and wildlife biologists and environmental justice scientists who work collaboratively with frontline communities and often come from those communities themselves. As environmental justice activists incorporate sciences into intergenerational struggles for civil rights and justice in relation to the land, they have theorized toxins as harming human bodies and the land simultaneously.[19] When they communicate scientific results in stories that connect people's own pain and difficulty breathing to animals' pain, illness, and death from shared toxic burdens, these affects move across bodies, drawing us into closer relation for a moment. Feeling for one another, we gain energy to recommit to struggle with a long horizon.

Field Methods: Autohistoria: Tell a Salmon Your Troubles

> The depressive position is a site of potentiality and not simply a breakdown of the self or the social fabric. Reparation is part of the depressive position; it signals a certain kind of hope. The depressive position is a tolerance of the loss and guilt that underlies the subject's sense of self—which is to say that it does not avoid or wish away loss and guilt. It is a position in which the subject negotiates reality.
>
> —JOSÉ ESTEBAN MUÑOZ, "Feeling Brown, Feeling Down"

This story will show one working of the trans-queer-underflow method I have been tracing so far in this chapter, by describing affects of relation that I brought from the field into a performance called *Tell a Salmon Your Troubles*. Let me tell you how that piece came to be, and how I later

Figure 2.2. *Tell a Salmon Your Troubles*, participatory installation at the 4S annual meeting, 2014. Photo by July Hazard.

understood its politics through Muñoz's writing on the depressive position in Nao Bustamante's *Neapolitan*.

I couldn't turn to ecology and a life of the mind until I came out as trans to scientists. The *trans* I mean begins with queerness and transgender experience but then extends to transspecies affections and transgressing disciplinary boundaries. These trans ways bring the field, grief, mourning, and joy at more-than-human animacy into sterile conference spaces. I got the idea during transits back and forth to the field during deep drought in early fall 2014. As the rain held off, I checked my study pools again and found more dead fish. The pools began to stagnate, and they stank of rot. Powerful farmers in California's Central Valley wrote op-eds in favor of salmon and smelt going extinct, so that farmers could divert the last few trickles of water that remained in the rivers.

When I did find fish that were alive in the few remaining shallow pools at my study site, I would talk to them. In the lab, scanning Twitter with then-postdoc Mike Bogan, I wondered what other scientists—and especially other queer ones—would say to these salmon, if they thought they could communicate. And then I wondered what would happen if scientists could speak their love and grief to their beloved species, in view of one another so not in secret, yet in intimate relation to that other? What would they say, and how would the saying affect them?

With my art-science collective the Water Underground as a sounding board, I conceived a salmon oracle in slapdash queer punk style. July Hazard and I first staged *Tell a Salmon* at the Bay Delta Science Conference in Sacramento, California, in October 2014. The performance took place at a folding table laid with a dozen-odd small round rocks. Pens and paper were available for participants to write down their troubles. July and I alternated in the roles of salmon and greeter. The greeter invited scientists and other passersby to sit down with the salmon, explaining that like all salmon, it wouldn't talk but could respond by turning over rocks. The salmon, wearing glittery formal wear and a giant spray-painted papier-mâché salmon head, sat behind a table making swimming motions. When a passerby sat down, she turned over one of sixteen stones arrayed on the table to reveal prompts such as "Describe a place you love that no longer exists," "Are you troubled by your data?" "What would you trade to save a species from extinction?" and "Do you like to swim?" Scientists responded in writing or by confessing their grief, confusion, and feelings of inadequacy to the salmon out loud. When the salmon moved its head as if swimming or turned over another stone, many participants interpreted these gestures as a response. Of the four hundred ecologists, hydrologists, and watershed managers who attended the conference, forty spoke or wrote to the salmon. Several spent more than thirty minutes doing so. The salmon appeared again at Glitter Bomb, a queer art show at SomARTS Gallery in San Francisco in June 2015. We then took the responses from both shows to the Cosumnes River, where we delivered them to the salmon by reading, by playing voice recordings underwater, and by dragging written messages behind our canoe. We performed *Tell a Salmon*, accompanied by recordings, written responses, and the video of our Cosumnes trip, at the inaugural Making and Doing session of the Society for the Social Studies of Science conference in fall 2015, where we argued that the approach was a Science and Technology Studies method for transforming ecological science in a queer-trans mode.

Theoretical Approaches: Queer Social Science, Trans Ecology

Can radical queer-trans-feminist analysis contribute to movements that extend solidarity across species boundaries by elaborating a critique of human-nature binaries?[20] To see fish and other creatures as making demands based on difference that will not assimilate to settler river control is one such move that expands the concerns of queer ecological inquiry.

The social scientist in me wants to know: did the paper salmon affect the other humans? In asking this question as one of the animators of the paper salmon, I am no longer just a human social scientist. The scientist inside the fish has been transformed and now must know what is happening to the other scientists and queers who cocreate the performance. At times, I felt the dead baby salmon from my field site swimming through the paper salmon to open a space for the scientists to feel grief. If they felt that grief, did that experience make space for them to change anything about their practice of science?

I envision answering these questions by conducting a series of interviews as the paper salmon, either live on Twitter, or in a marathon live performance at the next Bay Delta Science Conference, using a whole new set of questions set on the stones.

Preliminary thoughts toward further stone-questions:

- What organism(s) or ecological process(es) do you study?
- What makes your study organism cute?
- What makes it lovable?
- How would you describe your feelings for your study organism?
- Do you love it?
- Do you feel a sense of kinship with it?
- Do you think individuals of this species are dying a premature death?
- Who or what is responsible for this?

- Does it need to be saved?
- Who must save it?
- Do they love it?
- If you could tell a member of this species about your worries or fears for your future relations, what would you say?
- Do you ever speak out loud or write to this creature?
- Has it ever responded? How?
- Is your love for this organism erotic?
- Are you out to your friends and family about this love?
- Are you out to your scientific colleagues?
- If you confessed this love, would you be transgressing any boundaries?
- Do your study methods interrogate your response to the salmon and its response to you?
- (If no to previous) What would your next study looked like if you considered the salmon's response and yours to the salmon?
- What would you risk with this kind of study?

And

- Are you queer?
- Are you trans?
- Are you a scientist?

These questions raise theoretical questions that transgress river ecology and ethnographic norms. Our method elicited participants' responses to the paper salmon and living salmon's responses to participants' words we dipped and dragged in the river. How could another species respond to such an affective human gesture? How, indeed, could one know if a fish or insect or lynx ever responded? The project also blurred the boundary between art and ethnography, leading to questions of how much privacy and anonymity scientists would need to speak or write freely and whether we could share those words with other humans without jeopardizing the intimacy of that space.

Can scientists, through face-to-face interactions with the paper salmon, take up a depressive position by speaking their grief intimately in public? To explore this question, I now read *Tell a Salmon Your Troubles*—a queer-trans(species) performance—through Muñoz's queer reading of

Nao Bustamante's performance *Neapolitan*. Muñoz, discussing Bustamante's video/crochet installation, describes her decision to embody prolonged, repetitive, perhaps unending (yet endurable) grief on video as a taking up a depressive position and theorizes this decision as a queer-of-color move.[21] For Muñoz, this depressive position is potentially liberatory: Bustamante's performance of grief makes a space for gallery viewers of her installation to open to their grief. The scientists who told the paper salmon their troubles performed a public engagement similar to that of Bustamante's audience, talking earnestly and solemnly to the fish while their colleagues discussed data or grants over coffee nearby. Others refused this engagement, often expressing discomfort with the performance's camp aesthetic. A few commented that their interactions with the salmon led to new reflections on their data, their methods, and their stakes in multispecies relation. These scientists who talked to the paper salmon opened up a new psychological space in which they could reconcile their feelings of love and grief with their circumscribed political engagement. In so doing, they performed queer aesthetics and practiced trans(gressive) approaches to data interpretation. Following Muñoz, their "grief [was] temporally conjoined to ideas." The scientists and others who turned away from the salmon, or joked nervously about trouble with their mortgages, refused a transgressive and queer gesture. They didn't open up to transspecies feeling.

Field Methods: Autohistoria: On Writing as a Scientist

Back in Berkeley in 2015, as I finished my dissertation, I gave talks on the multispecies commons and the hyporheic imaginary that I saw emerging from these new configurations and modes of thought. My advisers and fellow students seem to accept my argument that this collaborative approach, deeply transgressive of boundaries between scientific disciplines and local knowledges, has produced good social and ecological science. Yet in writing up the academic papers from the research, for both social and ecological journals, I feel that I am excising the life, the animacy, the spirited feeling of openness and possibility from those field encounters, and the transgressiveness of the collaboration.

When I wrote the first draft of this chapter in 2016, I had just finished the ninth round of revisions on an ecology manuscript coming out of my dissertation work. The manuscript is an elegant thing, much improved by the keen minds and razor-sharp writing of my coauthors and focused by comments from reviewers: these supreme gestures of academic love and generosity. The paper is smooth and streamlined like a concrete canal, its inflows and outflows carefully monitored as in a water conveyance aqueduct, its seepages and blowouts and backwaters all straightened and confined. I chafed at some of these excisions over the last year of preparing the manuscript, but when I read it over now, I see there's no place in it for the missing material. The paper has a clear logic, the figures tell the story, it balances gracefully on its architecture of citations. Like the sleek aqueducts that crisscross Interstate Highway 5, it carries knowledge and results efficiently to other scientists but misses out on the messy intermingling in the oxbows and branching channels of a real river. Here, writing about other scientists' suppressed or compartmentalized grief and love as I read my own paper that replicates this move, I see how a queer or trans ecology is not yet here in my own scientific work.

The question, "What is there *not* room for in a science paper?" or its variant, "What lines have you not been able to cross?" is the space of inquiry in queer trans ecology. This is the space of the unknown, for the experiment—for the inhuman other of the experimental apparatus to speak. I've tried to write into this space in this chapter, using autohistoria. This experiment is at the interface of feminist/queer/trans critique and the practice of science. So far, I've sketched my methods; in conclusion, I offer a few results.

Conclusion: Results: What Is Missing?

In conclusion, in the absence of a queer-trans-feminist ecological subjectivity, an eco-hydrology manuscript takes on one or more of the following characteristics:[22]

- Speculation is excised explicitly throughout the paper and limited to narrow extrapolations of results in the discussion section.
- Politics and social dynamics may be mentioned briefly as motivation for the scientific project at hand but are not considered a part or product of science.

- Methods and modes of thought from humanities and social sciences are excised.
- Indigenous knowledge and land management practices may be mentioned only if they are written up in an academic paper in a quantitative field, which is then cited.
- Reflexive or auto-ethnographic material on the presumed relations between the scientific author and fish, landscape, and other human inhabitants, especially affective dimensions of those relations, must be avoided completely.
- The scientific researcher is not, in papers, supposed to describe peer relations with nonacademics, except in the acknowledgments to thank them for granting access or labor. The scientist is supposed to be an outsider to the community, whether human (for social sciences) or ecological. The practice, the learned styles and structures and syntax of science writing are supposed to discipline unruly relations that trouble this outsider status.

(Of course, everyone knows that these relations happen. Researchers fall in love with their subjects, get drunk with them, swim and sometimes sleep with them, are surprised by them, grieve for them. Everyone knows the best work at the scientific meeting happens after hours, at the bar. But scientific writing behaves in public as if this is not tacit, common knowledge.)

When July Hazard reads this section of my draft, he draws connections between this disciplining of the watershed through extractive cuts and scars and how the bodies of field scientists are cut with grief, joy, and perception. He suggests that the body of science has been similarly disciplined, the expansive thinking of natural philosophers channeled into narrow disciplines. These different forms of discipline, July argues, replicate the logic of the fish hatchery. The hatchery responds to the edicts of productivity and efficiency that is also the mode of the neoliberal university. The fish move from egg tray to tank to raceway, fed homogenous food, swarming to the surface with the same feeding strategy, protected from predators by a little net, fattened up as fast as possible, and then left to fend for themselves. In a hatchery, there is no seepage in the concrete tanks. The fish have nowhere to hide, and there are no backwaters or stagnant pools that push the evolutionary limits of what that species can become.

In the field, a scientist's body and psyche open to the other as the breach in the horizon where the world is ripped open. And yet this is not how we scientists write about it in the ecological science literature. That literature remains channelized within disciplinary levees. If a beaver-shaped meadow—with its blowouts and backwaters and uncontained seepage—is more productive of fish and other aquatic creatures than an engineered river, is the same true of an undisciplined mode of thought? A beavery mode of thinking, a transfigured body of literature, would let ideas seep back and forth, saturating the subsurface, flowing along complex paths and fostering new modes of production at the interstices of knowledges. This is the ecology of the undercommons, deemphasizing speed to completion or volume of papers produced in favor of a proliferation of circumstantial, located, embodied, meandering ways of thinking and dealing with knowledge. It may be that such an unruly ecosystem can hold more knowledge-code in its complex fractal of edges, just as an undercut and logjammed river holds more fish along the bank.

Consider autohistorias as queer ephemera, in conversation with José Muñoz—this time his writing on queer dancer Fred Herko's work "as both a choreography of surplus and a choreography of minor movements." Muñoz describes Herko's legendary last performance as "enacting a 'preappearance' in the world of another mode of being that is not yet here": a queer mode that refigures death and suicide as cultural labor.[23] Muñoz evokes Herko's movement and way of working through a scant archive, "as utopian traces of other ways of moving in the world [akin to] Negri's nuanced description of surplus value as an uncontrollable and potentially disruptive integer within late capitalism's formulations."[24]

Herko's final performance—a perfect jeté out the window of his East Village apartment, witnessed by his friend—comes out of the inescapability of death, of premature death, of the everydayness of it: Muñoz invokes all his friends who have taken their own lives. Then he himself leaves too soon, in what his friend Joshua Chambers-Letson describes as a too ordinary queer-of-color death. The queer performance Muñoz describes often pauses on the unsurvivability of queer life.

Tell a Salmon joins the precarious performance of out queer-trans affects within scientific conferences to the precarious position of salmon in California rivers in a time of relentless capital accumulation and disavowal of communal relations with riverine species. Salmon scientists at the Bay-Delta Science Conference watch the papier-mâché figure of the salmon perform movements that they have witnessed again and again in the field: whirling up as if in ecstasy, gracefully swimming, and also flopping in distress. Real salmon flop when they find themselves out of water. The temporality of near-extinction has a queer urgency dating back to Queer Nation manifestos in the late 1980s and was evoked by Muñoz's own friends and colleagues after his premature death in 2013. In 2020, COVID-19 made this repeated urgency, grief, rage, and overwhelm an affect shared by all those touched by the disease, again, mostly Black, Brown, and Indigenous people. To wait, to persist without any assurance of survival, to act with care in hopeless hopefulness, this is the shared situation of ACT UP, Black Lives Matter, Water Protectors, and border activists, who share traits of persistence and generosity with salmon.

July Hazard drew out this association in his 2007 play, *The Gold Fish, or, Straight Flushes for the Manifestly Destined* by chronicling a spawning Salmon who must play cards in a crooked casino to win her way upstream.[25] The Salmon—in the same papier-mâché mask and tattered finery we wore in *Tell a Salmon*, was performed alternately by July Hazard and Annie Danger. The Salmon flops frantically when she is dealt a bad hand by the Nymph dealer. The Salmon laments the unsurvivability of that swimming life, proclaiming, "My spawning grounds are silted! My fry will never hatch! Will never experience the thrill of ascending a waterfall—all the waterfalls have been submerged under *dams*! Oh, my fry, oh, my smolts!" The leap to radical politics comes in the final musical number, which opens with the Salmon making smooth swimming movements by weaving her head back and forth. She sings to the tune of "Do You Hear the People Sing?" from *Les Misérables*,

> Will you bless me on my run?
> Will you keep faith with fish and stream?
> When you stare at the rising sun
> do you dream our common dream?

The Salmon calls for solidarity, face-to-face in sharing of grief and rage that generates collectivity. This feeling is grounds for a public environmental science politics that goes beyond the doom and gloom of traditional climate change and environmental activism. Hazard's lyric

Figure 2.3. The Water Nymph faces off with the Tycoon as the Salmon and members of the Army Chorus of Engineers look on. Performers, left to right: July Hazard, Leah Morrison, Rori Rohlfs, Qilo Matzen, José Navarete. Photo by Erica Rubinstein.

invokes invitational animacy, a call-and-response from human to the other than human that rejects Manifest Destiny logics of river control. In response, one of the Army Chorus of Engineers, a human now gone over to the Riparian Dispossessed, responds,

> Do you hear the oceans roar?
> Hear the waves beat on the shoal?
> It is the music of the elements no human can control.

The Salmon, Engineer, and other human and nonhuman creatures of the Riparian Dispossessed Choir at last sing together,

> Will you swim and swim and swim
> until you cannot see the shore?
> Will you trust the wild whim
> of the storms outside your door?

In singing, we gain strength to keep swimming against unsurvivable odds.

Theorizing a queer x trans x ecology practice is excess to academic production within the field of ecology; such practice poses questions that transgress disciplines, resisting assimilation into any of them. Queer moves, then, resist colonizing impulses of academia that incorporate undisciplined edges into new fields of knowledge. Instead, by asserting this queer excess as undisciplined, we can stay partly on the outside, in the undercommons.[26]

This practice of resisting assimilation and refusing disciplinary bounds is not new for Indigenous scientists, wildlife managers, and activists. As Kyle Powys Whyte (Potawatomi) writes, Indigenous peoples have centered interspecies relations as cultural practices that are crucial to survival and have challenged power relations that privilege European knowledge and power relations in environmental governance.[27] In the Pacific Northwest, tribal and intertribal fish and wildlife science have transformed law, science, and policy around Pacific salmon and other species. Some non-Native fishers and scientists have joined these political struggles, mobilizing arguments of environmental injustice as they organize on many fronts: opposing natural gas terminals, removing dams, setting health standards for mercury in fish, and demanding industrial cleanup. In these struggles, Native discourse often emphasizes connection and relation: mercury moving through the food web harms salmon bodies and human bodies; ocean acidification destroys life possibilities for oysters and shellfishers; livelihoods connected to ecological processes foster cultures of stewardship that resist Manifest Destiny's extractive logics. Ecologists and other field scientists recognize more-than-human relationships and interconnections as real—ecology is, after all, the study of emergent properties of ecosystems—but mostly don't see them as cultural, as a social responsibility and ethical framework. Moreover, they rarely consider colonial legacies as determinants of ecological processes, or work to dismantle power relations as part of ecological work. If queer-trans-feminist field scientists join Indigenous and environmental justice scientists at the party and other settler-descendant scientists come, what wild moves they might make!

Underflow 3

Trans Thought as Latent to Manifest Destiny Logics

Consider how trans bodies are like bodies of water. Bodies of water are not containable. They overflow their banks in floods, or they seep away and go dry: the boundary is always a matter of contingency, and human choices on where to draw a line on a map. A river or lake is not contained by its shoreline—it seeps, hidden underground, underflows creating bogs or oxbow ponds that are at once discrete entities and tied by the water table to one other, so that they swell and sink together. Engineers now understand these connections but often find it more tractable to consider a river as water contained in a channel, separate from the groundwater, and to represent water as extractable and therefore separate from the life forms it contains and nurtures. This separation between water and river plays out when managers allocate water between "competing uses"—irrigation, municipal use, hydropower, and in-stream flows for fisheries.

In river management, this approach has been disastrous for fishes and other river denizens, whose habitats have been flooded for dams or dried out by levees, and for the fishing communities that depend on healthy fish populations. But it has also left some farmers and rural residents cut off from water in dry years, when water flows to cities and farmers with "senior" rights and groundwater flows to those with the deepest wells and strongest pumps. Dams and river control schemes are Manifest Destiny projects, and settler-colonial rationales are Manifest Destiny's lingering presence and process—ecological, socio-scientific, and psychological—within US river management.[1]

"Latent destinies" are counterparts to the Manifest ones that are unanticipated, often unrecognized, and in parts opposite in their tendencies. July Hazard and I have, across several joint and solo projects, explored resistance to Manifest Destiny as a latent and emergent shadow-side of US imperialism that can enact a Manifest Reversal by crossing back and forth across an ever-shifting frontier. This frontier has never closed or disappeared but rather shifts its locations and remains a powerful operator in US politics, romances, aggressions, and imaginaries.

Subsurface but not invisible, this self-reinventing frontier is a rich site for exploring Indigenous and settler legacies and for imagining their overturning.[2] As latent destinies, underflows are decolonizing and unsettling potentials that resist Manifest Destiny, both as a specific historical moment and as a continuing tendency of thought and action that is sedimented into infrastructures and technologies of governance. To think against or athwart the Manifest Destiny logics that created water infrastructures and regulatory regimes, July Hazard and I extended Eva Hayward's theorization of trans embodiment, which "reject[s] a binary between fragmentation and wholeness."[3] If we think of a river as a watershed body, then a trans river embodiment can refuse the utilitarian logic that dewaters and fragments a river and dries up its aquifer, so that the river holds no salmon and farmers' and farmworkers' livelihoods dry up. This trans river embodiment also refuses Edenic ecological imaginaries, such as those that some "beaver believers" hold. Beavers are not ecosaviors who can magically restore a right ecology or an unfragmented landscape. Beavers will make "novel ecosystems," just as invasive species do.

What concepts of trans(gender/species) embodiment bring to the process of hydrological embodiment—in bodies of water that hold the possibility of renewed ecological possibility—is attention to how the vicissitudes of settler legacy and present human need shape the water body and transform the water body and, together with the various species and elementals in the watershed, shape the possible presents and futures of that water body. We resist figuring the world as an already gravely wounded thing that humans must heal or make whole or as a dirty polluted thing that humans must purify. This refusal follows Hayward's rejection of the "simplistic" suggestion that "the pre-operative transsexual feels constrained by the 'wrong body' and longs to acquire the whole or healed body, which is represented by the male or female form."[4]

Rather, for Hazard and me, living a trans life while doing trans and queer theory feels inseparable from transgression. Figures of cutting/cutting across/crossing boundaries are resonantly transy figures, and putting ourselves in situations in which we are forced to cross binaries and boundaries often characterizes trans circumstance. Being trans means that we are always, on some level, transgressing these boundaries just by existing and are always betraying in order not to betray. This, Hazard and I propose, is the core sense of "a transing." Queer and queering are different from what we mean; queer is commonly understood
as a disrupting, slanting, skewing. This is not that, but the kind of irrevocable positioning that came out of Hayward's cuts and folds and that required the cuts and folds and the regeneration. This positioning knows itself in a sort of vertiginous joy at being unable either to turn back or to continue as before; new ways, transways, transselves are required of us here.

But it is also important that trans and queer not be contained or constrained within any discourse's demands for or expectations of productivity; once they can be deployed to orderly ends, queer and trans are no longer themselves and they are no longer ours—here "us" reaches to invite all queer and trans people.

3

The Watershed Body

Trans and Queer Moves in Beaver Collaboration

> The plan [in the undercommons] is to invent the means in a common experiment launched from any kitchen, any back porch, any basement, any hall, any park bench, any improvised party, every night. This ongoing experiment with the informal, carried out by and on the means of social reproduction, as the to come of the forms of life, is what we mean by planning; planning in the undercommons is not an activity, not fishing or dancing or teaching or loving, but the ceaseless experiment with the futurial presence of the forms of life that make such activities possible.
>
> —STEFANO HARNEY AND FRED MOTEN,
> *The Undercommons: Fugitive Planning & Black Study*

> The body, trans or not, is not a clear, coherent and positive integrity. The important distinction is not the hierarchical, binary one between wrong body and right body, or between fragmentation and wholeness. It is rather a question of discerning multiple and continually varying interactions among what can be defined indifferently as coherent transformation, decentered certainty, or limited possibility.
>
> —EVA HAYWARD, "Lessons from a Starfish"

> Trans theory's hyperawareness of categories and boundaries, its recognition of their instability and unenforceability, help trans animal thinkers like us explore interdependence, coercion, and reinvention . . . and undercuts the stability of resource-extraction logic with trans-species environmental imaginaries.
>
> —CLEO WÖLFLE HAZARD AND JULY OSKAR COLE,
> "Transfiguring the Anthropocene: Stochastic
> Re-imaginings of Human Beaver Worlds"

Introduction

The beavers' forests were burning all around us the first summer July Hazard and I met relocated beavers up close. That was in 2014, when engineers in drought-stricken California proposed massive dams and tunnels to store and regulate ever more river flow to fields and urban users. Subverting that logic, the biologists we met in eastern Washington relocate beavers to high mountain streams where they have been absent for nearly two centuries. They hope that beavers will build small, leaky dams not subject to human control. With time, these dams will create pools in floodplain meadows that shelter juvenile salmon (*Oncorhynchus spp.*) and release cool water downstream in the late summer. Driving down the Methow Valley before dawn, through dense smoke illuminated by flames on nearby slopes, we wondered what effects these beavers' future water projects might have on fire regimes.

Methods developed by US Forest Service biologists for live-trapping and relocating families of beaver on the Methow River have traveled to other rivers throughout the Pacific Northwest.[1] In trying to reverse the damage of twentieth-century river engineering projects, human actors are entering into improvisatory riparian relations with this animal that they often call an ecosystem engineer.[2]

Hazard and I thought with these beavers about how regenerative cuts can open possibilities of transformation for a human engineer who yearns against the totalizing assumptions of the anthropocene.[3] These assumptions—that settlers should control rivers, land, and earth process no matter the cost to Indigenous survivance, that technofixes can compensate for rampant extractivism even as climate emergencies deepen—we called an ongoing sign of Manifest Destiny's persistence. Native nations' long-standing and adaptable management practices fold cultural resurgence into caring for the land when they prioritize cultural focal species for restoration and center fishing, gathering, burning, and farming practices as management tools; the Tulalip Tribes' and Yakama Nation's beaver reintroduction projects exemplify this commitment.[4]

Decolonizing ethics challenges Manifest Destiny thinking, which assumes that only (white, male, straight) humans have ethics and that settlers should impose those ethics on Indigenous peoples and on the land itself. Beavers, Hazard and I argue, work athwart Manifest Destiny

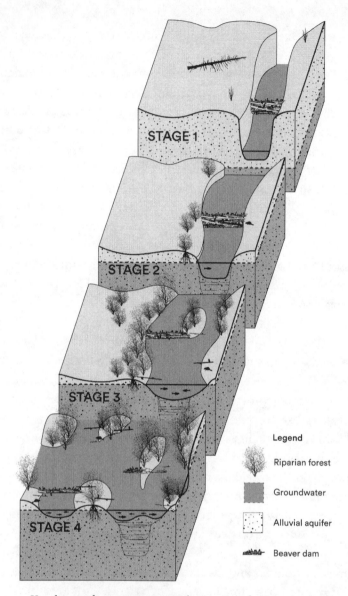

Figure 3.1. How beaver dams restore incised streams and raise water tables
Stage 1. In response to beaver removal and grazing, streams incise down into the floodplain, and floods rarely flow across it to recharge groundwater. Riparian trees die when their roots can't reach the water table. Stage 2. Beavers build dams that slowly trap sediment and raise the level of the stream. Willows re-establish. Stage 3. The water table rises in tandem. Riparian forest extends outward from the stream. Stage 4. Eventually, the channel rises up to the level of the old floodplain, and the stream may switch between multiple channels. Beavers sometimes abandon dams, which blow out, and build others. Figure by Morgan Southall.

imperatives: "Although beavers' ethics, if they exist, are not known to us, their physical undertakings on the continent transgress the territorial insistences and totalizing practices of Manifest Destiny. Their activity can thus bring about a physical decolonization; it also ties river systems and species back together in ways that increase resilience in the face of devastation."[5] Beavers are, at times, co-opted into Manifest Destiny projects; at other times, thinking with beavers about transgressing settler property regimes and river regulation can lead settler river workers to take up decolonizing ethics. Exploring cases where people enlist beavers to help restore salmon runs, I ask (1) what kinds of restoration projects may be read as transformations of the watershed body; (2) whether such projects disrupt or rewrite settler-colonial laws, properties regimes, and legal orders; and (3) what new ecological or ecocultural relations become possible as a result of these interventions.

Experimenting with trans thought as an underflow in river science and theorizing from trans experiences of bodily transfiguration, I probe the limits of analogy between trans and watershed embodiment. Trans theory disrupts settler purity discourses from an underground stream that shares traces of thought and action with Indigenous and postcolonial studies. Queer-trans-feminist relational methodologies, working in solidarity with Indigenous resurgence, can challenge settler property and legal orders—along irrigation ditches and in sloughs behind dams, by humans working not alone but in collaboration with other beings. This mode of river science and practice is open to ongoing reshaping. This constant flux and movement is inspired by rivers' mutually reshaping flows of water and sediment in dynamic interaction with plants and animals. A river—as watershed body—becomes more itself when these processes are not cut off or constrained by levees and dams. More itself might mean more salmony, more willowy, more wet, or even more flashy—a hydrological term for quick changes in flow that evokes a queer dance party. This transfiguring watershed body, I propose in conversation with Eva Hayward, is like a trans person who elects hormonal, surgical, or DIY transition. Thinking by analogy, such a land-water body might become more itself by means of bulldozer scars and beaver clear-cuts that might seem violent, unnecessary, or surprising to someone who doesn't see how flooding or stream-bank scour can trigger ecological renewal, just as many nontrans people see our body modification as perverse and damaging. In advancing this argument, I also explore how gender-deviant forms of expression that exceed transnormative discourses of wrongness and rightness might apply to rivers. What rivery futures might emerge if river lovers of all genders rejected individualistic discourses of bodily integrity and reconsidered the watershed as a kind of body-in-transition? I hope that this trans analysis inspires river workers to stop trying to create an ideal, "right" land or water or ocean body and, in the process, to work to transfigure river restoration practice to make space for genderqueer or nonconforming human embodiment.

To explore these questions, I build on my and July Hazard's speculation that taking up trans politics of embodiment might lead river engineers to embrace change, surprise, and unexpected flourishing on scarred but vibrant rivers.[6] By making a transspecies alliance with beavers, we argue, engineers might "translate physical transsexual survival techniques into psychological trans-species survival invitations," finding some way to follow the practice of Hayward's trans figures who "create embodiment by not jumping *out* of our bodies, but by taking up a fold in our bodies, by folding (or cutting) ourselves, and creating a transformative scar of ourselves."[7] This embodied trans politics is a spirit of always-crossing that can animate river governance through reciprocal responsibility. I apply this politics to the field engagements of restoration workers who work with beavers to improve salmon habitat. I elucidate a trans approach for trans ecologists and for others who work for watershed recovery but don't yet ally with trans thought.

The chapter roughly follows a trajectory from abstraction to embodiment. I first visit beaverless Salmon Creek—where a sense of "yearning against" settler river modification and trapping fever inspires utopian rewilding campaigns. I then travel to the Methow and Klamath valleys, where people grapple daily with beavers' furry bodies and gnawing antics, and solicit them as partners in repair. Finally, I visit Seiad Creek on the Klamath River, where beavers' transgressive acts—stealing willows from a restoration project to do the work themselves—led a restoration worker to reimagine his work in the watershed. So that metaphors of watershed embodiment and transfiguration remain grounded in trans people's lives, I traffic back and forth between rivers and trans and gender-deviant performance as sites of queer affinity with and beyond the human. In the conclusion, I identify decolonizing ethics that sometimes emerge in the contact zone where queer, trans, and Indigenous theory meet.

Seeking Ghosts of Lost Beavers

In my Salmon Creek field site on the central California coast, beavers were trapped out in the 1800s and have still not found their way back.[8] Coho salmon and steelhead trout thrive in the river-bottom wetlands that beavers make wherever they go. Coho are nearly extinct across California, and struggling in the Pacific Northwest, in part because most of these habitats have been diked and leveed for agriculture, transportation, and urban sprawl.[9] Near the mouth of Salmon Creek, to bring coho salmon back from the brink of extinction, biologists release several hundred adult coho salmon from the Warm Springs conservation hatchery each winter. These "broodstock" fish have lived their whole lives in freshwater tanks. They are genetic outcrossings of the last few hundred fish remaining in the Russian River coho population in 2001—a spawning population estimated at one million in the late 1800s. Biologists—including my mentor, Michael Fawcett—removed these fish as fry from shrinking pools in drying tributaries in 2001 and have since rebuilt the run to a few thousand spawners per year.[10] Local school kids carry the silver fish, ripe with eggs and milt, from tanker trucks to the stream in nets and release them to wait for the winter rains, then spawn up in the tributaries where they will.

I studied two of these tributaries, Fay Creek and Tannery Creek, to figure out where young coho and steelhead fry holed up during the long summers and whether the creeks held good habitat and enough water for their populations to recover.[11] The answer was, in the drought years of my study, that more than 90 percent of the fry perished; the population will not recover under these conditions. In wet years, enough fry do survive that populations will increase if conditions in the estuaries and ocean are also good. But given that climate change is making drought years the norm, people need to improve habitat—and increase streamflow—if these fishes are to recover. As I walked the cow-trampled dry channel each August and watched pools shrink and young fish disappear, I imagined beavers building massive dams that let the deep, cool pools of June persist through the summer. I felt the lack of water—and of beavers—as a bodily sensation, a dry parched heat and yearning for that lost water's cool weight, which pushes it down into the voids between riverbed stones.

I recognized this feeling of loss and yearning from the stories of other stream ecologists—friends and colleagues I'd invite to snorkel the creek with me, counting fish. In most years, coho and steelhead teemed in the clear, cool water, especially in a wide, gravelly reach of Fay Creek where deep pools formed. Then, in July or even late June, the trickle of water across gravel riffles slowed, then stopped, leaving disconnected pools with long reaches of dry streambed in between. I walked the two creeks week after week, measuring the water's temperature and oxygen content and the size of each pool. By the first rain in October, on average, half of those pools had dried up. Others were black and fetid with rotting leaves. Deprived of oxygen, the few fish that survived barely moved to conserve energy. Their gills flared red as they struggled to breathe.

What those young fish experienced during drought—how they struggled, suffocated, and died prematurely—results from settler refusal to care that is rooted in misrecognition. Settlers misrecognized how redwood trees fed summer fog into soil and how logjams and beaver dams recharged winter floods so that water would trickle into the stream in August. Settler loggers and ranchers have transformed forests and grasslands directly by plowing and logging and indirectly through fire suppression. Since the 1880s, local Pomo and Miwok people, recently rerecognized as the Federated Indians of Graton Rancheria, have not been able to practice cultural burning; the frequent, cool fires they once set increased summer streamflow by clearing brush, keeping evapotranspiration low; cultural burning also reduced the severity and frequency of wildfires. Traditional late-summer burns can also reduce summer stream temperatures at critical times for salmon by creating smoky inversions; further north on the Klamath River, Karuk and Yurok people set fires ceremonially in August to welcome—and trigger—salmon migration.[12]

Collectively, settler ways of using and managing these watersheds have left them desiccated, overgrown, bleeding fine sediment into streams, and prone to catastrophic fire. For years the county paid a local rancher to drive a bulldozer down the stream, scraping up logs that might back up floodwaters; by removing logjams and filling in pools, ranchers and conservation district workers removed salmon refuge habitat and also reduced the amount of floodwater that seeped back into floodplain aquifers.[13] Then, beginning in the 1970s, developers subdivided large ranches into residential parcels, each tapping its own well or spring into the shallow and fickle aquifers.

The Salmon Creek Watershed Council and I discussed what restoration techniques could help salmon survive this drought and adapt to climate change, which is predicted to make extremely dry periods like 2011 to 2015 the new normal. Beaver dams can buffer temperature extremes and rewet desiccated floodplains. That is why river restorationists in the blasted and dehydrated rivers of the US West are either recruiting beavers to dam streams or emulating beaver dams themselves, by weaving post-and-brush structures that retain water. Some Salmon Creek Watershed Council members wanted to bring these "beaver dam analogues" to the watershed, to add to their current tools of rebuilding logjams and recontouring stream channels. To decide which techniques to use, and where they would do the most good, the Watershed Council and I decided to map the tributaries each September, when streams are driest. Volunteers marked pools that stay wet, dry reaches, pools that hold salmon, and trickling springs that add enough oxygen that salmon can breathe. These oxygenated pools, our maps and analyses showed, were fed by groundwater.[14] If neighbors could keep groundwater flowing to the stream through the summer, more salmon might survive. Our maps and timelines of settler change and salmon population decline sparked conversation at the council's farmers market table, where volunteers explained that if people pumped less groundwater, salmon might survive and recover. They showed examples of ways to do this—by catching rainwater in tanks, thinning and burning the forest, and building ponds that would let more rain seep into and recharge local aquifers.

Lost Beavers in the Trace

When Darwin described bees' pollination of *Catasetum* orchids, he used performative language suggestive of an interpretive dance.[15] Looking back on these five seasons of collaborative salmon studies, I realize I and my collaborators were caught up in a similar, gestural call-and-response with other species.[16] When I waded through beaver ponds not yet flooded, under overgrown alders not yet chewed down, I imagined fat darting salmon smolts in August's black anoxic pools, which held only dead young fish.

When I saw traces of lost beavers in incised channels and dry wells, I tried to feel or sense the physical underflows of water beneath the ground. Thinking about these physical underflows attuned me to conceptual underflows: collective imaginaries that, if put into action, could transfigure settler science governance into acts of desire for vibrant river worlds. Curious if local scientists and residents saw the stream this way, I conducted formal interviews with local residents and agency scientists and kept ethnographic notes of field days with these collaborators. From these data, I used grounded theory methods to uncover different conceptual models of the watershed as a damaged body. I sketch those models here to speculate about how beavers might transfigure this shared imaginary.

One early summer day, I climbed up Fay Creek's narrow upper canyon with Jay, a creek-walker born on a local commune who keeps track of steelhead and coho spawning. His two boys roamed ahead and startled a family of mergansers. They were feeding on coho and steelhead fry in a deep pool below a basalt outcrop. The ten-year-old climbed up the black knob of stone and carried down a translucent goose egg shell. Only a thin trickle flowed between pools. With three more hot months to go until the fall rains, Jay was imagining beavers as the creek's salvation. "This could all be a beaver sanctuary," he said. "No one comes down here, and the rancher who owns it would be happy—his well went dry."

In fall, I crawled up main-stem Salmon Creek with Sierra Cantor, a local agency biologist, snorkeling to count young coho and steelhead. She told me that the rancher who owns this stretch of stream thought beavers had once built dams here and that they could help restore the stream. But Sierra thought that because most land was subdivided, regulators would block beaver reintroduction, fearing complaints from landowners. Brian Cluer, a NOAA geomorphologist, agreed: "It's going to take more than just landscaping existing channels—we need to be in the real estate business and restore floodplains and wetlands."

The next spring, local beaver advocate Kate Lundquist invited me to a community presentation on bringing back the beavers. Lundquist's presentation chronicled a six-year campaign to return beavers to coastal California streams. Interspersing photographs of happy beaver scientists thigh-deep in beaver ponds situated in an otherwise dry streambed, she showed Russian and Spanish records that enumerated local beaver pelts traded in the state and gave Pomo and Miwok words for beaver. With this evidence, she worked to convince California Department of Fish and Wildlife officials to change the current policy that blocks beaver translocations. But Lundquist also needed physical evidence to clinch the case that beavers are in fact native to the Coast Ranges, and thus translocations would represent a reintroduction of a native species. She asked the audience—mostly local residents—to keep an eye out for chewed sticks, layers of organic debris from old dams, or—best of all—beaver bones.[17] Talking of beavers, Kate moved her arms fluidly, mimicking their swooping dive under the water, mimed chewing sticks, and swept her arms wide to indicate the expanse of water they backed up behind a dam. Watching her performance of love for beavers and their unpredictable ways, some audience members seemed to catch Lundquist's "beaver fever," responding affectively with glimmerings of biocultural hope.[18] During the Q&A, one member schemed gleefully how beavers might show up on their properties, or in Willow Creek State Park.

However, the audience members also felt hopeless against apathy, drought, greed, and climate change. One said that her neighbor "is pretty hardcore 'don't come on my land.' He'd probably shoot the beavers." Private property regimes limit beavers' ability to work at scale to heal the landscape—to reshape processes to heal the watershed's wounds. Residents take walks through a small nature preserve's trails, delighting at a hooting owl or yipping coyotes, but most don't directly work for watershed repair. Even as they allow themselves to respond emotionally to nonhuman species, they delegate work in the watershed body to scientists and restoration practitioners. Without face-to-face engagement with beavers, salmon, or other multispecies kin who could lead them into a transfigured watershed body, the audience members felt overwhelmed at the prospect of climate change adaptation within current regulatory structures. In contrast to the biocultural hope elicited by beavers' gestures—and Kate's transmission of them—the affect elicited by the beaverless watershed is one of dissociation from the watershed body because of alienation from other watershed inhabitants. In contrast to residents, professional biologists and restoration workers work in the watershed day after day. They often delight in multispecies encounters in the field yet excise their feelings from professional communications. This dissociation from the affective and the aesthetic limits their projects' potential to effect political transformation.

Queerer, trans strategies—ones that seek reembodiment in relation to the more-than-human: this is the salmon- and beaver-inspired practice I call coming into a transfigured watershed body. Beavers dam prolifically and then move on. They give water away with no thought for scarcity. They cultivate forests to make abundant and dynamic habitats for other creatures, just by being beavers: chewing and building and living their lives. Salmon swim all over the sea, storing up nitrogen and trace elements in their bodies, then leap countless obstacles to swim them far upstream. They leave thousands of eggs and their very bodies to feed people, bears, raccoons, skunks, and eventually trees.

In settler-minded landscapes like Salmon Creek, changing settler laws and property regimes to allow beavers to do their transfiguring work will require that people there transfigure their relationships to the watershed body. Kate Lundquist's talk began that transfiguration when she presented ephemera of beaver encounters. These photos and chewed sticks sparked excitement and anticipatory joy in her audience of settlers dismayed at drought and impending salmon extinction. Now let us delve deeper into this trans feeling of yearning against settler-colonial affects that linger in river ecologists after encounters with beavers or their sign. Even encounters with these ephemera as traces—in Kate's stories and gestures—infused some of her audience with a desire for beavers' profligacy that transgressed wet-dry boundaries and settler property lines.

To explore this yearning-against as a queer and trans feeling, I turn to José Esteban Muñoz's writing on the trace in relation to Black gender-deviant dancer Kevin Aviance, who sometimes performs in New York nightclubs frequented by cis, gay "gym queens," most of them white. Though this leap may be incongruous for some readers, to me this exploration of gesture and its residue captures this yearning-against exactly: "The hermeneutics of residue on which I have called are calibrated to read Aviance's gestures and know these moves as vast storehouses of queer history and futurity. We also must understand that after the gesture expires, its materiality has transformed into ephemera that are utterly necessary."[19]

Muñoz's passage theorizes from gesture's transits between human bodies in a dance club. I take up this transit among human bodies in the field and between human and animal bodies when we notice, startle, or attract one another. Gesture is embodied movement. So, gesture can be a shared language of human and other-than-human persons—an expression of recognition and feeling that subtends and exceeds spoken language. Muñoz theorizes gesture as a circuit of feeling through a collective of lovers and strangers, and it is this aspect of relationality that is crucial to the argument I am making about circuits of trans-bodily affect within a collective watershed body. In "Gesture, Ephemera, and Queer Feeling: Approaching Kevin Aviance,"[20] Muñoz describes how Aviance's drag performances elicit a circuit of feeling between himself and his performers that subtly challenges white supremacist tendencies of mainstream gay culture—and, implicitly, the homophobia and transphobia and anti-Blackness of dominant US culture:

> I invoke the phrase "approaching Aviance" because I want to cast a picture from life, the scene of Aviance's being approached. To travel through the gay world of New York City with Kevin Aviance is certainly to call attention to oneself. Aviance is six foot two, bald, black, and effeminate. In or out of his unique drag he is immediately recognizable to anyone who has seen his show. To walk the cityscape with him is to watch as strangers approach him and remark on one of his performances. They often gush enthusiastically and convey how much a particular performance or his body of performances means to them. One will hear such things as "I'll always remember that one show you did before they shut the Palladium down" or "You turned it out at Roxy last week." Kevin will be gracious and give back the love he has just received.
>
> His work, his singing and his movement, is not the high art of Bill T. Jones or Mark Morris, but I would venture to say that more queer people see Aviance move than have witnessed Jones's masterful productions. I do not mean to undermine the value of Jones's work. I only want to properly frame the way in which Aviance's nightlife performances matter. The gestures he performs matter worlds to the children who compose his audiences. Aviance is something of a beacon that displays and channels worlds of queer pain and pleasure. In his moves we see the suffering of being a gender outlaw, one who lives outside the dictates of heteronormativity. Furthermore, another story about being black in a predominantly white-supremacist gay world ruminates beneath his gestures. Some of his other gestures transmit and amplify the pleasures of queerness, the joys of gender dissidence,

of willfully making one's own way against the stream of a crushing heteronormative tide.

The strong influence of vogueing practice in his moves affirms the racialized ontology of the pier queen, a personage who is degraded in New York City's aboveground gay culture. Often, one gesture will contain both positive and negative polarities simultaneously, because the pleasure and pain of queerness are not a strict binary. The conversations that ensue after his performances, the friends and strangers that approach him on the street, the ads in bar rags, the reviews in local papers, the occasional home-video documentation, and the hazy and often drug-tinged memories that remain after the actual live performances are the queer ephemera, that transmutation of the performance energy, that also function as a beacon for queer possibility and survival.[21]

I do not mean to compare Kevin Aviance's performances to a beaver's insouciant displays of trespass against settler property and river control regimes. Rather, I want to explore the affects that both Aviance and beavers bring into being among their fans. Aviance's fans "gush enthusiastically and convey how much a particular performance or his body of performances means to them"; beaver believers tell stories of beaver performances, over and over again, to make sense of the unruly interventions that beavers make into river restoration. I propose that responses that pass between beavers and their human admirers are queer feelings. As queer affects of the inhuman circulate, they trouble human-nature boundaries. When people mimic beavers' ways of building dams, engineering is no longer solely a human act. Thinking with Muñoz about the human and beaver gestures such engineering engenders, this associational method considers my and others' accounts of encounters with beavers as queer traces that can spark trans ecological thought. This associational method is inspired by Muñoz's influential 1996 text, "Ephemera as Evidence":

Central to performance scholarship is a queer impulse that intends to discuss an object whose ontology, in its inability to "count" as a proper "proof," is profoundly queer. The notion of queer acts that this opening essay hopes to offer is immediately linked to a belief in the performative as an intellectual and discursive occasion for a queer worldmaking project. Thus, I want to propose queerness as a possibility, a sense of self-knowing, a mode of sociality and relationality. Queerness is often transmitted covertly. This has everything to do with the fact that leaving too much of a trace has often meant that the queer subject has left herself open for attack. Instead of being clearly available as visible evidence, queerness has instead existed as innuendo, gossip, fleeting moments, and performances that are meant to be interacted with by those within its epistemological sphere—while evaporating at the touch of those who would eliminate queer possibility. Tony Just's images are emblematic of the kind of invisible evidence which I will proceed to unpack as "ephemera." Queer acts, like queer performances, and various performances of queerness, stand as evidence of queer lives, powers, and possibilities.[22]

In my project, ephemera as evidence shows how beavers make river engineering into queer trans acts. Key to this argument are my own transits between queer DIY performance and river restoration activism worlds. Before returning to Muñoz on Aviance, I describe some embodied encounters—with humans and beavers—from these worlds. Queer trans traces of beavers touch Muñoz's realm—the NYC queer performance scene—through networks of queer affiliation among me, Hazard, the Water Underground art-science collective we conspire with, and our sometimes-collaborators, the performance artists J. Dellecave and Ezra Berkley Nepon.

On tour for our 2007 anthology *Dam Nation: Dispatches from the Water Underground*, July Hazard and I dreamed up a fantastical casino as a way to convey the high stakes of water politics and how settler-colonial institutions stack the deck against river creatures; this sketch became July Hazard's play *The Gold Fish, or, Straight Flushes for the Manifestly Destined*. J. Dellecave cocreated the Water Nymph character—an out-of-work river guardian with PTSD from the water wars—and played the Nymph in stage performances and the film adaptation.[23] The Nymph moonlights in the Gold Fish casino, run by a tap-dancing Tycoon and featuring the Army Chorus of Engineers as the house band. The Nymph deals cards to a Salmon; each card becomes an obstacle she must face on her migration upstream. At the bar, members of the Riparian Dispossessed Choir lament their changed circumstances, including a Beaver who—every time the Salmon laments a human-made dam—asks eagerly, "Dam? Somebody call for a dam?"

Two other performance projects reembody these beavers, amplifying gestures and extending beavers' performance of unpredictable dynamism—or stochasticity—into queer performance spaces. J. Dellecave's installation *Nocturnal Beaver* is, in a sense, a Muñozian trace—an ephemeral intensification of feeling that borrows from scientific concepts, then transfigures them via collective queer-trans production.[24] Dellecave described the installation as "an unpredictable disruption of flow similar to how a beaver dam disrupts the flow of water."[25] Dellecave brought the papier-mâché beaver head and tail from our film set to NYC and wore it during a three-night endurance performance at the MIX festival. She gathered cast-off objects from artists and (mostly gay male) partygoers into an "accumulating sculpture" hung on an armature of broken tree limbs collected from New York City streets in the wake of Hurricane Sandy. The endurance installation evoked queer- and trans-inclusive projects and a "flagrant pants-off celebration of dyke sexuality" simultaneously.[26] The storm's energy mingled with the late-night sexual energy of the party to create a shifting, dynamic collective embodiment of transgression, not least by Dellecave's femme nudity in the predominantly cis gay male space.

In Ezra Berkley Nepon's Purim *shpil No One Mourns the Wicked: The Wizard of Shushan*, Nepon wrote a speech for the Witch against White Supremacy, who, dressed as a beaver, declaimed a queer, trans, beavery sense of stochasticity: "It means the randomness that can emerge from complex systems, like storm and flood patterns, or social movements! The same processes occurring in the same environments do not always lead to the same outcomes! In chaos, anything is possible. When beaver dams break, as they so often do, beavers quickly adapt, rebuild, or move on. Beavers are stochastic stars, dancing gracefully with the winds of change!"[27] This sense of beavers as working against white supremacy in its guise as anthropocene engineering amplifies and expands Hazard's and my theorization of beavers as trans figures.[28]

Reader, consider these queer reworkings of stochasticity as "storehouses of queer history and futurity" for queer and trans ecologists and artists.[29] Consider them, simultaneously, as "ephemera that are utterly necessary" for heterosexual and cisgender river workers who are enacting profound interventions through river restoration.[30] The mainstream, surface flow of ecological restoration is rooted in Western science and often aims to return ecological communities to a state of wholeness. This "reference state" is usually an Edenic, idealized landscape, in which Native people are either absent or making little mark on land, river, and sea. Restoration practitioners, both Native and non-Native, have spilled much ink in the pages of ecological restoration journals on how to identify the "right" reference state and on ethics that rest on the purity of the resulting ecosystem.

But understanding stochasticity as a willful and exuberant practice of nonhuman actors working in concert shows that this approach is limited. Surface flow restoration ethics emphasize purity: a lack of visible human engineering and invasive species. Settler fantasies of an unspoiled wilderness tended by "noble savages" permeate ecological restoration literatures, including in concepts like "rewilding." Underflows to mainstream restoration—and especially Indigenous, queer, trans, and feminist approaches—challenge these settler-colonial fantasies, with evidence from Indigenous science and management but also, even more crucially, with different ways of knowing and sensing ravaged landscapes and using feelings of despair and heartbreak to envision and gesture toward repair. These approaches work in the trace, by seeking the intensification of dynamic feeling that come from being together in the field and telling stories of ecocultural pasts and futures. Underflows approaches refuse purity as an aim or endpoint. They do not reject nonnative species outright but rather incorporate them into ecocultural practice: we delight in the Himalayan blackberries that ripen every year, while experimenting with different ways to keep them in check. Such underflows approaches take direction from Native nations' and place-based communities' collective ethics of reciprocity and work to support their ongoing management work.[31]

Toward Embodiment: Touching Problem Beavers

In the trace, let us now think with Muñoz about how handling and moving problem beavers from place to watery place might be akin to queer performance that evokes "feeling the joys of gender dissidence."[32]

The first time I touched a living beaver was during the summer of 2014, on a Columbia River tributary, during an era of extinctions, loss, shock, reassessment, and disorientation that is coming to be called the anthropocene. I traveled there with July Hazard, continuing years of scientific, artistic, and political engagement with beaver-salmon-human worlds. But only now did we engage physically with the immediate and lively heft, squirm, and odor of *Castor canadensis* in the flesh. With biologists from the Washington Department of Fish and Wildlife who were working with the Yakama Nation and the US Forest Service, we helped relocate "problem" beavers to territories where people welcomed them, with all their unpredictable land and water interventions. On a ranch outside Ellensburg, we piled logs and brush into a crook in the stream, making a "starter lodge" where newly released beaver pairs could hide out and get their bearings. After, we drank lemonade on the rancher's porch; she spoke of exploring the streams as a young girl and eager anticipation of beavers' reworking of those streams. Then we drove to a square pond where beavers had made a nuisance for county workers by plugging culverts and blocking water flow. We set two chain-link traps just below the waterline, camouflaged with willow branches, on either side of a square cage holding a baby beaver—lure for the mother and cousins, who were not fooled by artificial scent. We added some paper towels soaked with castoreum from other family members for good measure.[33] The next day at dawn, we found one trap empty and one full of a sturdy young male beaver. Hearing us, his massive mother poked her head up in the middle of the pond, gave us a long, piercing stare, then dived and SLAPPED! the water with her tail.

The beavers we met made excessive, flamboyant gestures, not only to the human workers who attempted to trick them into traps but to the imaginaries of settler control of water in territories long claimed conquered by Manifest Destiny. Beavers' works inevitably transgress private property lines. Beavers interrupt human irrigation schemes and chew down orchard trees. Their dams transform landscapes from fence-able pastures and drained fields into dynamic mosaics of thicket, meadow, water, and bog. When beavers absconded from human-chosen release sites in the Methow River headwaters, some traveled up to fifty miles downstream, while others built dams and lodges in agricultural areas and foraged on commercial apple trees. A few disappeared from the biologists' surveillance network of microchip PIT (passive integrated transponder) tag-readers.

Beavers flout boundaries between the human and the natural, and between land and water, and inspire people who collaborate with beavers

Figure 3.2. Methow Beaver Project staff trap beaver families to move them higher in the watershed. Photo by Alexa Whipple.

to restore rivers to do the same.[34] This mixing of human labor and beaver life has its own ethics, a beavery trans ethics some scientists and passionate amateurs take up when they protect beavers' dams in city parks, or drive wooden posts behind beaver dams so they can withstand floods.[35] These acts reject settler wildlife, forest, and river management ethics that are similar to the ethics of plantation science. Anna Tsing writes of plantation science ethics, "Experts and objects [such as beavers] are separated by the will to power; love does not flow between expert and object. . . . For those who love wild mushrooms, such control is not an object; indeterminacy is part of the point. Wherever volunteers gather to promote disturbance forests, or pickers stop to ponder why the mushrooms come up, plantation science loses a little authority."[36] The feeling of plantation science is an old feeling, steeped in settler nostalgia for control. In some settler wildlife managers steeped in plantation science ethics, beavers engender feelings of frustration when human control over rivers slips. Like the matsutake mushroom tenders Tsing describes, who trouble the economic logics of forest plantations in Japan and Oregon,[37] these beaver abettors trouble the hydrological imperatives of large hydroelectric and irrigation systems when they embrace beavers' stochastic ways.

Along the Methow and Yakima Rivers (tributaries to the Columbia), as in the Russian River, where Kate Lundquist works, people who collaborated with beavers felt such old desires to control floods or beavers but also shared new feelings: a love for beavers that embraces indeterminacy and transformation and a fuck you to the authority of the dam schemer and the irrigation engineer. These feelings can inspire trans ethics. I call them "watershed feeling" and "dissident feeling," respectively.

A few days after our encounters on the Yakima River, July Hazard and I helped different biologists retrieve live traps—like a chain-link suitcase packed with a heavy, toothy beaver—at dawn from sloughs and ponds along the Columbia River near the Methow River confluence. In the Methow Valley, orchards and forest plantations have overrun beaver habitat, and trappers have long thwarted beavers' recolonization. Now, climate change is reducing snowpack and leaving rivers warm and nearly dry by late summer, while wildfires scorch vast areas, triggering landslides and burying salmon spawning streams. Sometimes the only areas of green left across a burned forest are beaver pond complexes, which create firebreaks and shelter wildlife from the inferno. Recently, a statewide ban on lethal leg traps created an opportunity for live trappers to relocate "nuisance" beavers from agricultural lowlands.

We followed the pickup full of beavers back up the Methow to a repurposed fish hatchery that serves as a holding facility. The beavers stay here for a few days while the biologists try to trap their other family members, often by baiting the trap with the beaver's actual scent, or even with a beaver's own baby. They then release them into streams high in the mountains on National Forest land.

At the hatchery, we helped the biologists wrestle each beaver into a Kevlar cone until just its nose and tail stuck out. Then, we hoisted it onto a scale, injected a microchip PIT tag into its tail, plucked a few hairs for genetic analysis, attached an ear tag, and finally reached up into its anus to squeeze some oily castoreum from its scent gland. The biologists all smelled this fluid and discussed whether it smelled more like motor oil (indicating a male beaver) or blue cheese (female). They then gently lowered the Kevlar cone into a concrete pond and let the beaver waddle into a makeshift plywood lodge. Once released into the old salmon-rearing tanks, the beavers hid in makeshift lodges, then ventured out, to choose partners promiscuously or refuse them, to fuck or fight or hole up and chew willows while always looking for a chance to build a dam. If adolescent beavers shacked up with new mates, the biologists released them together, into the starter lodges that biologists built from logs and willow branches.

Touching, wrestling, and smelling these beavers, July and I attuned to the feelings that resonated and traveled among the people who encountered them. One old feeling, expressed by some ranchers, farmers, and foresters, is that "beavers are a real pain in the ass." Beavers and their works are messy and disrupt human plans and labor. People and beavers come to know each other through a call-and-response of building and disrupting one another's works. Beavers build dams across road culverts and irrigation ditches; people tear them out. Beavers chew down orchard or ornamental trees; people shake their fists and replant. The old ethics, which still persist, led settlers to respond to beaver depredations by shooting or trapping the beavers and destroying their lodges. They did so to protect improvements to their private property and extract value from land that they separated from floods and drained of groundwater.

I want to consider floods as flamboyant excess—not just of physical water, but of ecological generosity engendered by wasting (water, nutrients, carbon) as selfless abundance. Excess, following Muñoz, is a relative term: "It is not so much that the Latina/o affect performance is so excessive, but that the affective performance of normative whiteness is minimalist to the point of emotional impoverishment. Whiteness claims affective normativity and neutrality, but for that fantasy to remain in place one must only view it from the vantage point of US cultural and political hegemony."[38]

River management, from the vantage point of US cultural and political hegemony, has long figured floodplains and wetlands as wastelands, excess to settler productivity through agriculture or industrial development. This hydrologic excess transgresses and scours away settler property. I want to refigure this excess via Muñoz's theorization of Latinx affects as not too much, but rather appropriate, and revealing normative affects of whiteness as a lack of feeling. Encountering lush beaver- and salmon-rich floodplains that Native people managed for food and fiber, settlers exalted Protestant repression of strong emotion and actions of domination and control. They claimed divine right to overrun sovereign Native nations and called it Manifest Destiny. They denigrated Native peoples' affects, and relational practices of reciprocity to other-than-humans in ceremony, as indolent or animalistic. As miners dredged and blasted away river bars and hillsides, ranchers drained and plowed floodplain wetlands, and industrialists dammed, straightened, and confined rivers, these settlers asserted that control and dissociation from more-than-human worlds as commonsense and productive. These men were not all white, yet they performed a cisheteronormative, white, masculine affect. Like the Latinx affects that reveal the impoverishment of the affective performance of normative whiteness, river workers' trans affects of excessive feeling for beavers, salmon, and other river beings reveal white settler affects as lacking, because they stifle response and reciprocal action.

One such trans affect, *watershed feeling*, is at play when settler farmers learn about beavers' engineering work and decide to call the state biologists to translocate "problem" beavers rather than killing them.[39] If these people form an affective kinship with the beavers through feelings of excitement, anticipation, or relief at the abundance of water and wildlife that beavers attract, such transspecies affinities may work against the old feelings engendered by separation, domination, and control. For example, one biologist talked with admiration about a beaver who swam more than fifty miles from a release site, only to swim back and set up a dam nearby, while still damning the beavers for the trouble they caused him. On a ranch in the Yakima Valley where we built a starter lodge to house relocated beavers, he told the rancher, "The beavers have been hiding out, avoiding the traps. They're wily and persistent. They'll swim miles to find the place they want to be, and they'll figure out a way to dam." He said this with excitement, his eyes lit up, and seemed to feel lucky to witness and aid beavers in hanging on as people trap them, neglect their willows, and breach their dams.

These feelings modulate the old feeling of exasperation at loss of control. One rancher, after telling the story of his decades of unsuccessful efforts to stop beavers from damming a stream and flooding, said, "Now we just love to come down and watch them and think about what they have been building and pruning." He and his wife said they still think beavers are a pain in the ass, but they have decided that, in the balance, beavers' good qualities outweigh the bad, and they have made peace with the flooding. (They control how deep the water gets by inserting a drainage pipe, called a pond leveler, partway up the dam.) As people form personal relationships with beavers, some even say they love them. When beavers moved into a flood control channel in downtown Martinez, California, local activists named "their" beavers and tracked the kits' birth and growth and blogged about their antics in language that grandparents often use for their grandchildren. Kent Woodruff, a Methow Forest Service biologist, described on film his feelings of lack and grief when they find beaver dams burned up in wildfires, leaving ponds drained dry.[40]

A second trans affect, *dissident feeling*, is at play when beavers' transgressions against property boundaries and river control infrastructure inspire people to embrace and emulate these practices in order to become better commoners in multispecies worlds. Beavers flaunt their disdain for property lines and human infrastructure by blocking flows and flooding land. Among beaver believers, I often heard that beavers look at a culvert and see a great dam with a hole in it that humans forgot to plug up; one state wildlife biologist said this wistfully, as if wishing he could let them do it. Beavers remind some settlers that by welcoming floods and adapting to drought, they can share water profligately and embrace stochasticity that creates life chances for nonhumans. Beavers are not engineers working for humans, and they don't conform to white settler heteronormativity. Michael Pollock, who executed influential studies of how beavers invigorate degraded streams, said that river sciences' obsession with river control blinds settler river workers to effective Indigenous strategies for renewal.[41] Settlers who first blocked the river from flooding roads and fields (farmers, ranchers, engineers) and those who have begun to resist such river control (river scientists, restorationists) relish stories of "problem" beavers' wily transgressions of settler river control logics. For Pollock, corralling beavers as unpaid pond-builders makes it harder to see what beavers are doing: they are constantly making promiscuous relations across species lines and reintroducing unpredictability that fosters abundance.

What politics flow from naming as trans affects beaver believers' feelings of transgression and refusal? A key innovation of trans life has been to assert a multiplicity of ways to do gender and to emphasize collective action and performance as strategies for transforming not just individualized trans selves but also collective spaces of possibility for attacking binary strictures and remaking them into a dazzling array of gender presentations and embodiments. This transgression of a singular body extends, for example, to trans protocols around pronouns and bathroom signage, which rest on the understanding that an individual person is constituted relationally with others. To extend this act of transgression and refusal to a watershed body—comprising relations among humans, nonhumans, and landscapes—human engineers might reject affects of shame at being in the wrong watershed body, engineered to exclude beavers and salmon, by transgressing and disrupting engineering norms with illicit beaver reintroductions and other rebellions against stodgy regulatory processes.

In narrating their gender transition, many trans people emphasize self-actualization in an individualistic sense. Dominant US narratives glorify individual expression. Ideas of queer-trans "chosen family" also emphasize individual choice and mobility to cities with queer scenes. Ecological systems, however, aren't chosen networks—everyone works together, in a sense, not because they chose to but because they are there. This larger ecology of trans community co-creation includes the tolerance for myriad ways of doing gender, which is akin, perhaps, to beaver lovers' tolerance for the beaver's sometimes inconvenient activities. For me, and some others, it is important counter these individualistic narratives with experiences of trans becoming through a sense of mutual recognition that one's "self" was never singular. Like coming into one's trans body through trans community cocreation, becoming a member of a transfigured watershed body requires that one always act in ways that foster good relations. Grounded in the experience of coming into one's own trans body with others, in an ongoing process, we can think now about how river workers might coming into a transfigured watershed body.

In interviews and field conversations, I learned that rural settler landowners who welcomed beavers sometimes did so in order to act as a part of this watershed body and to feel connection that exceeded their property boundary. In the Scott, Methow, Seiad, and other agricultural valleys, farmers often constrain the riparian corridor with levees or neglect it until the stream threatens to flood their fields. Work with beavers offers a chance to join in with those flows that exceed the property boundary, to temper the urge to control where streams flow, and to transfigure the relationships among land, water, and species. Water comes in and leaves by the ditch or the well but also seeps through the substrate, floods over the surface, and flows past in the stream. Through beaver collaboration, river workers and beaver lovers imagine and materialize connection via water's surface and subsurface flows. A trans underflow way of doing this is to ask, What watershed body is most joyous, bountiful, and most vibrantly itself, given its past histories, its current politics, and the future desires of the people who hash out research protocols and fish habitat designs in a specific riverine place? In relation with humans and other-than-humans, such trans moves challenge the fixity of categories.

Beaver Gestures in the Trace

Having described some bodily and gestural encounters between humans and beavers, I now turn to a more focused consideration of these gestures: as ephemera, in the Muñozian sense, that make a queer kind of evidence. In 2007, on an island in Kwakwa̱ka̱'wakw territory, Hazard and I paddled a canoe through a narrow strait from a small lake to a big one and docked at a massive cedar log that had been floating so long that ferns and huckleberry bushes sprouted from its mossy, decomposing bark. We counted thirty-four bald eagles perched on snags around the lake, screeching and jostling for the best perch. I remember the water was still as glass. I don't remember exactly what we talked about—the land, the water, the eagles, our speculations on histories and ongoing lifeways. I remember we talked easily there, in the way of queer trans lovers who have driven many back roads to many minor lakes and paddled around them for a while before stopping to think about Manifest Destiny and its possible reversal. At some point in that afternoon that seemed to stretch to fill all of time, a beaver head appeared right in front of us, followed by a brown humped body on a fast vector for the eagle shore. We watched its wake spread across the water. One of us said something, and the beaver twisted its body up in the air, fixed its beady eye on us, SLAPPED! its tail on the water, submerged, and came up chittering.

Another time, Hazard and I hiked six miles to a wilderness lake on the western fringes of Glacier National Park, in Kootenai and Kalispel territory. The trail ended at a campground, which was completely covered with pine and fir—normally avoided by beavers—felled by chew marks so that they crossed the tent sites and made the campground totally uninhabitable by humans. Across the lake, a lone beaver lodge poked up through the water.

It was sometime after returning from Glacier that Hazard and I wrote these notes, now evoked as a trace, a reminder of the origin both of words and of chapters: "'Trans' is meant to disturb purification processes" and "'queering has the job of undoing 'normal' categories.' Plunging these insights back into beaver waters, we see beavers displaying themselves wonderfully in this light, never working for human ends; never deployable to build a particular dam or recharge a particular aquifer or respect any property line."[42] These encounters and responses are, in some sense, queer and gender-deviant encounters that generate queer ephemera.[43] Beavers, however, do not know human histories or politics and cannot intervene in them with studied bodily gesture or denounce policy from a dance-stage-turned-pulpit, as Kevin Aviance does.

Evoking what José Esteban Muñoz does for queer gestures on the dance floor, I turn now to an exploration of interspecies gestures along streams—how they can allow people to reimagine the past as a different set of not-yet-here potentials. To show what I mean, I invite you to come on a transit from a gay New York dance club to a beavery slough in Karuk territory on the Klamath River. The limits of this comparison become evident as soon as I try to describe beaver gestures. Unlike Aviance's gestures on the dance floor, beaver gestures are not transmitting and amplifying the pleasures of queerness or the joys of gender dissidence. But if we understand settler river control as a central force of heteropatriarchy, then human scientists' pleasure at witnessing beaver-driven river stochasticity could be seen as queer-trans disruption of settler-colonial policies—if those pleasures disrupt settler fantasies of control. When a river worker like Will Harling, who will come into this story below, exults to find beavers stealing his stockpiled willows to build a dam where he'd planned to build a brush weir, might he be expressing joy in river dissidence that rejects the settler norm of the "productive" and "fully appropriated" stream?

Beavers help such river workers make queer feelings and trans moves that disrupt heteropatriarchal fantasies of Manifest Destiny. What I want to call attention to here is the feelings that beaver gestures engender in humans who are (or who yearn to be) river outlaws, ones who live outside the dictates of settler river-normativity. Muñoz writes, "In [Aviance's] moves we see the suffering of being a gender outlaw, one who lives outside the dictates of heteronormativity." To be against heteronormativity is to make queer-trans-drag moves. Aviance's moves disrupt settler colonialism in its guise as white, cis, male homonormativity on the dance floor. It's not that beavers themselves make a trans watershed body by digging canals that divert waters out into a meadow or by abandoning a dam so that a spring freshet can breach it, scattering stored sediments and turning pond back into stream. We can't know what beavers feel or why they gesture the way they do. However, Aviance's *drag gestures* and the beavers' *dragging willows gestures* converge in my mind, "willfully making one's own way against the stream of a crushing heteronormative tide," by means of a holographic image kindled by Muñoz's tidal-river metaphor.

Though this queer-trans reading alongside Muñoz reveals crucial resistance against settler colonialism and white supremacy as manifested in river management, there are pitfalls in thinking in parallel. Aviance is making gestures of pain and pleasure as a conscious conversation and artistic intervention into his audience's worlds. People who danced below him at clubs respond to Aviance in the street, and these feelings and affects flow additionally through exchanges of thought and spoken words, which engender and transmit ethics. The beavers' gestures are not primarily for their human observers in the same way, nor can beavers even be said to be making gestures so deliberately. For me, Muñoz's narration of Aviance's performance brought back memories of fabulous drag performers in clubs and warehouses and also of seeing beavers or their traces in the field on various tangled riverbanks. Queer gesture and queer embodied experience activate something—feelings, desires, yearnings against—that is a kind of Muñozian ephemera.

Just as Kevin Aviance's performance gestures expand affective possibilities for white, butch, gay men, beavers' river gestures expand affective possibilities for white, straight, cisgender river scientists. Muñoz writes,

Imagine the relief these gym queens feel as Aviance lets himself be both masculine and feminine, as his fabulous and strange gestures connote the worlds of queer suffering that these huddled men attempt to block out but cannot escape, and the pleasures of being swish and queeny that they cannot admit to in their quotidian lives. Furthermore, imagine that his performance is something that is instructive, that recodifies signs of abjection in mainstream queer spaces—blackness, femininity/effeminacy—and makes them not only desirable but something to be desired. Imagine how some of those men on the dance floor might come around to accepting and embracing the queer gesture through Aviance's exemplary performance.[44]

As strange as it is to imagine gym queens as mainstream ecologists, engineers, and stream restorationists with grown-out hippie haircuts, in goofy shoes, covered in muck and beaver castoreum, it gives this queer trans outcast great pleasure to do so. Reading the above passage, I imagined these river workers, whose river-shaping projects must always operate within constraints, coming around to accepting and embracing the queer gesture.

In most western US rivers, restoring fish habitat depends largely on private property owners' goodwill—as on Salmon Creek, where several ranchers collaborated with local agencies to store rainwater and replace logs in the stream channel. Restoration, including hyporheic and multispecies common imaginaries, is often thwarted by settler colonialism, which is reenlivened through strategies of river control: water law artificially separates ground and surface water; levees and mine tailings separate rivers from their floodplains; dams change streamflow and block fish passage. Some settler river restorationists admire beavers' flagrant disregard for such constraints. When these people exult in the salmon and other species that throng in beaver ponds like masses of gay gym boys dancing at the feet of Kevin Aviance, are they "accepting and embracing [the beaver's transgressive] gesture"? If they then collaborate with beavers to make transgressive hydrologies, how might such deviance and transgression rework science? Muñoz continues, "More important, imagine what his performance means to those on the margins of the crowd, those who have not devoted their lives to daily gym visits and this hypermasculine ideal, those whose race or appearance does not conform to rigid schematics of what might be hot. Those on the margins can get extreme pleasure in seeing Aviance rise from the muscled masses, elevated and luminous."[45]

I am one who would be with Muñoz on the margins of the club. As a transmasculine ecologist who often passes, I also often find myself on the margins of straight men's misogynist chatter in cars on the way to the field. This may be why beavers' deviant gestures took on such suggestive meaning for me. I came to study science from the margins, being queer and trans and having studied in the streets and the landscape instead of finishing my bachelor's degree. To stay in science in graduate school, I needed to queer and trans it—by doing undisciplined thinking and performance alongside formal social and natural science research, with the artist-scientist collective the Water Underground.[46] Muñoz writes, "For queers, the gesture and its aftermath, the ephemeral trace, matter more than many traditional modes of evidencing lives and politics." Thinking with and through the trace is, especially, a queer method—and as Muñoz makes clear, often queer in a transy, gender-deviant mode. Writing in this mode can sharpen the politics of river restoration and suggest new contexts and approaches for science and restoration practice. To close this section, I show how, using evidence from our beaver encounters and performances as singing Riparian Dispossessed creatures, Hazard and I theorized beavers' transgressive lessons for an archetypal engineer who yearns against settler fantasies of river control:

> [Ecologists] understand beavers through the human construction of "ecosystem engineer" and ascribe all kinds of human qualities and values to them. A trans reading might understand beavers as transgressive, as disruptors of ecosystems and human works, whose own works yield not only biodiversity but human possibility. Beavers transgress against human concepts of engineering, and against human works, but not against the river. In speaking of beavers, speaking of "crossings" may generate possibilities that speaking of nonspecific "transings" does not. Etymologically, trans and cross share the meanings "across," "beyond," "traversing," and "on the opposite side." Crossing has other valences, too—to strike through text, to oppose or contradict someone. A beaver is adept at crossing—beavers fell trees across a river. Crossing might be a good way to think about beaver. A

beaver crosses, swims from one side to the other, and chews on things on both sides. We wonder whether shortening trans to a prefix, and habitually not specifying trans what, allows some of the elisions, imprecisions, disembodyings, and defusings we notice in some trans theory. What, if anything, might one say better or know better by keeping track of the different strands of what trans can be? Could sharper attention to the different trans valences or cadences help us locate more accurately power's vulnerabilities to transgressions?[47]

Unsettling Moves along the Klamath

Looking for evidence of such trans feelings and comings into the watershed body, I came across river practitioner Will Harling's writing about beavers, which conveyed some of those gestures and feelings and mobilized them to repair settler-colonial harm to the Klamath River and its ecocultural relations. In this final river visit of the chapter, I propose that a trans theory of watershed embodiment can sharpen and focus these narratives of unsettling projects. This theory can support settler river workers, like Harling, who work in solidarity with Indigenous tribes and practitioners, within a larger project of unsettling that returns land and management authority to Native nations.

Harling is a non-Karuk biologist who grew up in Karuk territory and learned about Karuk science and ecocultural practice in the Tribe's curriculum in the local schools. He went on to direct the Mid-Klamath Watershed Council, which works closely with the Karuk Tribe on salmon habitat restoration.[48] Like the Yakama Nation's watershed management projects, Karuk Department of Natural Resources ecocultural restoration projects amplify dynamic landscape processes. Protocols for monitoring and harvesting different species, and for managing them using cultural fire, respect other-than-human beings and their world-making. When Yakama, Karuk, and other Native watershed managers began to collaborate with beavers, they did not need to transfigure the degraded watershed body or theorize people as part of it—they already recognized that reciprocity, and they exercised their responsibility to the watershed body through ceremony and ongoing harvest and management. But for Harling and other settler river workers, beaver encounters upended and overturned their assumptions about the proper relationship between humans and other species. They came to recognize, through beaver gestures of dam building and fugitivity, a different role for human river workers than that of the anthropocene river engineer who yearns only to control water and messy ecological dynamics. By following beaver gestures in the trace, settlers like Harling, Michael Pollock, and Kent Woodruff reimagined their place within a transfigured watershed. Hazard and I theorized this process as a trans experience of embodiment—of coming into the watershed body together with other humans and other species.

Harling wrote a story about a beaver encounter and how it made him reimagine his river restoration work.[49] Reading Harling's story and talking with him later helped me think about how (nontrans) Indigenous theory and trans theory can make parallel paths toward river justice. Because too little trans theory engages Indigenous scholarship and protocol, I want to show how a trans reading of a settler narrative of a beaver encounter can work in solidarity with Native unsettling projects.

In his story, Harling tiptoes out of the house carrying his baby and a fishing pole, careful not to wake his wife, and walks to a good fishing spot. Of this place, he writes, "The relatively wide Orleans Valley gives the river a chance to meander a little here, reclaiming its sinuosity stolen over the past six million years as the Klamath Mountains began to rise from underneath, forcing it into steep sided canyons tracing fault lines in the uplifted bedrock just upstream and downstream of the valley." At the river, he encounters geology, water, fish, and, for the first time, a beaver:

> Across the river, I noticed a furry head moving slowly upstream. The light brown tuft of hair visible above the water looked like what I thought a beaver would look like, but couldn't be sure. Just then I heard a rustle of grass and a swish of a tail on the near shore and backed into the willows to watch. Sure enough, a beaver was swimming up towards us along the edge of the river just twenty feet away. . . . Big whiskers and a large black snout, those dark beady eyes and two cute little ears quickly disappeared when it spotted me, and a loud thwack of its tail as it dove alerted its kin that danger was near.

Harling sees something—is it a beaver? It is a fleeting sighting, but his biologist's eye for a combination of characteristics clinches it. He responds by backing away "into the willows." The beaver's response to Harling, "a loud thwack of its tail," brings him into relation with the beaver's kin, "alerted . . . that danger was near." The beaver is startled. Harling is giddy. That relation attunes him to other beaver signs and hungers as he walks home to breakfast: "Giddy with excitement from this rare close encounter, I noticed all the stripped willow sticks along the shore, even a clump of uneaten willow shoved under an algal mat, possibly left for a mid-day snack."

In this passage, Harling gives the river and other geologic features an animacy that, in English, is usually reserved for humans, especially white, settler, cismale, heterosexual ones.[50] The river "reclaim[s] its sinuosity" from the mountains that have "stolen" that sinuosity, "forcing it into steep sided canyons," and the salmon "navigate" up "through the shallows."[51] In speaking for and with the various riverine creatures and elementals in a multivocal chorus, Harling is taking up a new position within and of the watershed body, refusing the position of the disinterested, unaffected engineer who sees rivers as drainage channels and water as an inert and lifeless fluid. In his identification with the wily beaver who is hiding out and reworking mining-damaged floodplains beneath the notice of settler river authorities, Harling is taking up what Fred Moten and Stefano Harney have called the fugitive art of social life, but he is doing so within a sociality that includes other species as well as humans. Like the strategies Moten and Harney call for in *Undercommons*, this watershed politics is practiced "in animative and improvisatory decomposition of [politics'] inert body . . . [and] emerges as an ensemblic stand, a kinetic set of positions."[52] Harney and Moten name multiplicity and multivocality as key strategies in opposition to the single authoritative voice that characterizes state planning: "Its encoded noise is hidden in plain sight from the ones who refuse to see and hear—even while placing under constant surveillance—the thing whose repressive imitation they call for and are."[53] State planning, steeped in Manifest Destiny logics, misses the signal in the stochastic noise. Here Harling is seeing the signal—what ecologists gloss as emergent properties, like biodiversity—as of, and emerging through, the noise of the predawn watershed body: its gurgles and thwacks and tweet-tweet-tweets, its multiplicities of relations and responses unfolding backward and forward and in circles through time.

As he engages with the beaver and draws himself into affective relations with the floodplain world, Harling extends this affect of care to beavers that were trapped and shot relentlessly during the nineteenth and twentieth centuries:

> Beavers are slowly coming back to the Klamath, recovering from intense trapping that began in the mid-1800s and continuing for nearly a century after until they were almost extinct. In 1850 alone, famed frontiersman and trapper Stephen Meek and his party reportedly trapped 1,800 beaver out of Scott Valley, which at the time was called Beaver Valley. The last beavers in Scott Valley were trapped out by Frank C. Jordan in the winter of 1929–1930 on Marlahan Slough. Beaver throughout much of the Klamath basin suffered the same fate, and even today as they return to less inhabited areas along the mainstem river and its tributaries, they are still shot and trapped in streams where their dams pose a perceived risk to residential and agricultural property.[54]

In imagining beavers' experience of trapping, he emphasizes the beavers' own agency (the quick disappearance at the sight of the dangerous human, their foresight in stashing some willows to eat later) in their return to the Klamath as a sign or possibility of subversive human-beaver relations. This agency is key to the later part of the story, in which humans recognize beavers' improvisatory dam-building gestures as collaborations for watershed renewal.

Conclusion: Beaver Dreams

In bringing trans theories of embodiment back to beaver-salmon worlds, I want to ask, "Where or in whose imagination might this watershed body be emerging?" in order to more fully theorize the watershed body as an imaginary that is shaping science and watershed politics. Following Eva Hayward's "toward myself through myself," the transformation of watershed body into multispecies commons is not a change from one state or body into another.[55] Rather, it is the recognition of an already existing multispecies relation into which humans can enter. In the Klamath River story, I see this move already partly enacted in how Will Harling talks about the beaver he meets, as call-and-response, among beavers and humans who recognize one another as reshapers of their shared worlds. The collaborations on the Mid-Klamath work toward decolonizing, by centering the Karuk and Yurok ecocultural knowledge and contracting with the Tribes' natural resource departments to carry out and monitor restoration work, which is then amended and reworked by beavers.[56]

The Methow and Yakima beaver projects also evidence a move toward unsettling, with the Yakama Nation contributing land and personnel, the Forest Service lending equipment and expertise, and the Department of Fish and Wildlife trapping and relocating beavers. Here, too, beavers provided a way for trappers, ranchers, and hydrological scientists to reimagine and reenact key strategies of Manifest Destiny (trapping, diverting water, logging, forest management) in a transfigured and relational way. In Salmon Creek and the Russian River, which have been beaverless for at least 150 years, community scientists begin to imagine a transfigured watershed body through salmon surveys and wet-dry mapping. However, most Salmon Creek residents' watershed imaginaries do not include beavers' stochastic engineering, because California wildlife codes prohibit moving beavers from place to place; only advocates like Lundquist, who seek them out on other rivers, meet beavers face-to-face.

Thinking with beavers and rivers and mountains and salmon, out in the watershed commons, I have sketched a way to transfigure what anthropologist Anna Tsing calls "capitalist ruins" into substrates for new modes of life.[57] Practices of world-making with beavers and salmon unveil the multispecies commons that is always-already there, in the land, ready for human acknowledgment and recognition. This joyful response to the leaps and thwacks and dives that are beavers' or salmon's modes of communication mixes with human affects of loss and grief at their wanton killing and extinction.

Scientists and others who enter into this common watershed body describe feelings of joy and sensuous pleasure in fur and musk and slippery fins. Akin to queer feelings that Muñoz maps in punk clubs and on queer dance floors, touches and glances of recognition kindle visceral feelings of relation that are queer ways of making kin. Once that recognition takes place, does it trigger in human hearts and guts the feelings that we feel for human kin, queer or otherwise? Protectiveness, grief at death and loss, joy at birth and growth and promiscuity, empathy for past depredations: to recognize these as feelings of kinship would shift ethics. Maria Puig de la Bellacasa describes one such feminist ethics of care that arises from encounters with other species: "Because 'nothing comes without its world' we do not encounter single individuals, a meeting produces a world, changes the color of things, it diffracts more than it reflects, distorts the 'sacred image of the same'. Knowing is not about prediction and control but about remaining 'attentive to the unknown knocking at our door', But though we do not know in advance what world is knocking, inquiring into how we can care will be required in *how* we will relate to the new."[58]

I want to extend this thought on feminist care ethics to queer-trans practices of making kin, by thinking about how people might join beavers and salmon in their stochastic world-making. As people develop practices of caring for beavers and salmon, they necessarily transgress boundaries among institutions responsible for fish, wildlife, and water management. Their boundary crossings reveal these institutional divides to be artifacts of Manifest Destiny fantasies of dominion and control. Queer-trans ethics flip over those fantasies like spawning female salmon flip over gravel with their beating tails. The undersides of those fantasies reveal Indigenous knowledge and an ongoing practice of caring for whole networks of species through the cultural practices of burning, harvest, and ceremony.[59] These transgressions across land-water and species boundaries inspire people to engage with rivers in new ways. In new studies by Pollock, Woodruff, the Karuk Tribe, and ranchers in the Scott Valley, beavers have insinuated their world-making practices and flipped old Manifest Destiny questions. Rather than studying how to drain water away, how can we preserve hyporheic flows? Rather than trapping out a (generic) nuisance tree predator, how can we entice this particular beaver family to live here by matchmaking and enacting practices of interspecies care? Such studies enter into a call-and-response with beavers, salmon, forests, and weather, as scientists are inspired to improvise with beaver dam analogues, starter lodges, beaver deceivers, and pond levelers.

How these incipient collaborations will play out—for beavers and humans—is not settled. People who love beavers do not all talk of the present, past, or future in the same ways. But there are points of convergence.

To hold another's perspective in mind, to enter into being together in-difference, these have been key tactics in environmental justice and women-of-color and queer-of-color politics. In closing, I extend this solidarity to other watershed inhabitants by means of a queer-trans utopian gesture—the closing section of Sarna-Wojcicki's and my performance "The Manifest Reversals of Multi-species Collaborative Watershed Restoration".[60]

> *Dan Sarna-Wojcicki*: In entering into this multispecies collaboration, humans can join in transgressive ecologies that reengender symbiotic relations of becoming-with. The ensuing improvisations can break through political impasses and undo the cataloging impulses that underlie antagonistic human-nature constructs. These changing relationships between people and beavers, and accompanying legal and policy changes, may be signs that a Manifest Reversal is underway.
>
> *Cleo Wölfle Hazard*: We found the following prophecies scratched in mud banks by beaver claws and scribbled in the interstices of our interview transcripts:
>
>> Signs of latent destinies:
>>
>> Settlers acknowledge Indigenous sovereignty through comanagement and repatriation of ancestral lands, return of stolen artifacts, and protection of access to ceremonial sites.
>>
>> Humans and beavers develop collaborative practices, installing pond levelers and beaver deceivers. For every beaver deceiver, there is some trade in territory to beavers' advantage.
>>
>> Evidence of ontological shifts, such that humans no longer think they must dominate natural processes at the expense of nonhuman others.
>>
>> In recognition that human and nonhuman fates are bound up together, beavers are no longer seen as furbearers to be killed frivolously, but as collaborators in ensuring the world's continuing liveliness.
>>
>> When Manifest Reversal is at hand,
>>
>> Rivers will retake many of their floodplains. Rivers will no longer be fixed by riprap so that property lines may also remain fixed.
>>
>> Beavers will chew their way up all of their former streams and, where they encounter humans, will be aided with support

structures and accommodated with pond levelers but everywhere be permitted to grow enough willows that they can chew in peace.

Salmon will return in silver hordes. Their dead bodies will turn into giant trees, which will be permitted to grow to a great size. Where some are cut for human uses or put back into streams, they will be cut selectively and dragged out over the snow, one by one.

Such a reversal would be as large in scope as Manifest Destiny was, rolling out across the continent and transforming every river, aquifer, and floodplain.

Both voices: And the beaver SLAPs resound!

Underflow 4

Making Queer Kin and the Queer Field

Against assimilationist portrayals of well-adjusted middle-class white gays and lesbians, I understand queer as a space of refuge and political activism. This space is made and remade with the love and labor of trans people of color, gender deviants, disabled people, and other misfits who may have sex with people of a different gender but, through their relations with and in queer community, join queer kinship networks. The queer field science practice I imagine grows from practices of being at hand, caring for our elders and agemates and young ones together, reaching across time to touch our queer ancestors through writing. Sometimes we even find each other in the university, on Twitter (@500queerScientists), and at professional meetings (like GayGU at the American Geophysical Meeting). To think about how queer kin-making practices could build solidarity with Indigenous sciences, I turn to Shawn Wilson's *Research Is Ceremony*, which crucially centers relationality as guiding practice, epistemology, and ontology for Indigenous co-researchers in North American and Australian contexts.

Arguing that "relationships do not merely shape reality, they are reality," Wilson draws on collaborative study with other Indigenous co-researchers to outline research approaches that are accountable to relationships.[1] Grounded in Terry Tafoya's (Taos/Warm Springs) postulate that it is not possible to know the definition of an idea and the context for that idea simultaneously, Wilson's approach rejects the extraction of Indigenous research from its setting and proposes an alternative: bringing the reader into a circle of relationships and accountabilities.² He writes, "It is my intention to build a relationship between the readers of this story, myself as the storyteller and the ideas I present. This relationship needs to be formed in order for an understanding of an Indigenous research paradigm to develop. This paradigm must hold true to its principles of relationality and relational accountability."³

For Wilson, sharing context and building a relationship with his unknown reader is crucial to an Indigenous research paradigm. Beginning with letters to his three sons explaining his project, he moves among genres—academic prose, transcripts of conversations with collaborators—and eventually brings the reader into relation such that he can blend the different voices. Wilson's approach invites a reciprocal one, a multigenre and participatory approach to writing that can draw the reader in to my questions and methods. *Research Is Ceremony* also invites reciprocity in sharing family origins and understandings of kinship that make sense of blood and nonblood ties, of adoption and nonmonogamy but also childbirth and welcoming new human life and the relations that can be made with and through that child.⁴

Like Wilson's project, mine is one of cross-cultural communication, a search for common purpose across multiple axes of difference, among them natural science and the humanities, Indigenous and Western science, academic and activist spaces, and different lived experiences of race, class, gender, migration, and region among my collaborators on ecocultural restoration and water governance projects. His letter to his sons prompted my own reflection on how queer family-making practices shape the queer relational methodologies I am developing. Here's the letter I wrote to some of my queer kin:

> Queers hold birthdays sacred, and celebrate them with great extravagance. We also come together around other life transitions, other making of a new life—moving to a new town, graduating from nursing school or with a doctorate or from witch school. We come together around illness, around suicide and the detritus it strews among our remaining networks. We come together in the streets, against wars, against gentrification, against police shootings, with the longshore workers, for Palestine, with Standing Rock, and then again against police shootings, or in the streets, with can-

dles or just our own gorgeous, grieving bodies to mourn another murdered trans woman of color.

 Recently, more and more, we come together around the actual birth of a new child. In my case, this queer kinship started with who can physically make the embryo, DIY sperm counts, home science projects in various gamete configurations that eventually ended up with two people who never wanted the experience of reproductive labor (me) and who never wanted genetic descendants (my good friend and straight-identified former lover) combining gametes that (after much labor by me, my queer trans partner July Hazard, and support from assorted queer relations) become a child for the collective to raise. Rather than the futurity the straight version of family expects—the continuation of genetic heritage, the passing on of tradition—the queer family welcomed more the present unfolding: scouring the internet for a gestation app that wasn't horribly transphobic in its gendering of parent and fetus, texting me developmental trans species affinities that app provided every few weeks (your baby is as big as a pangolin! your baby is as big as a brioche!), the sometimes awful and sometimes hilarious search for butch pregnancy drag, many hands giving massage in the interminable underworld space of a 47 hour long home birth, and the everyday labors of bringing food and postpartum care that all families do.

 And then, with the kid not even a month old, and me still forbidden by my midwife to lift more than the weight of the baby, these queer kin packed up all our stuff into a moving van and bid us goodbye. A family photo shows that bittersweet moment of longing for and leaving our closest kin to move 800 miles north, battered green canoe strapped to the roof of the car, three hours late, smiling.

 In the best case, queer family redistributes reproductive labor far beyond the nuclear family, and extends the family beyond queer space defined by sexual relations (though there was plenty of sexual history in that family photo too). And yet the kind of labor our queer family did for us was not so different from what we all do for each other towards non-reproductive projects, to shepherd a film or land project into being, or to grieve during an illness or after a death. Rooted in punk DIY cultures, anti-capitalist commitments, collective work, and street protest, this kinship network extends beyond sexual cultures to welcome a wide range of misfits and outsiders, and to spread and deepen this culture through work together on projects.

As Shawn Wilson's Indigenous research methodologies are for and by Indigenous scholars, the underflow methodologies that I elaborate in the next chapter are for and by queers—including queer trans people and heterosexuals who have made relations with queer communities. It also is for resolutely transdisciplinary (and undisciplined) scientists who don't identify as queer but whose work can't be contained within the natural sciences, and in particular, those who work with grassroots and tribal communities. These different ways of not fitting in are necessary, but not sufficient, for an underflows methodology. What is necessary is to bring modes of practice and cultures of collaboration, learned in queer community, to our labs and field sites.

Field spaces offer temporary respites from the violence done by enforcing whiteness and gender norms, but those norms persist in labs, classrooms, and conference rooms. In the car on the way to the field, in lab meetings and courses and conferences, to not be seen as queer, trans, Native, Black, Latinx, working-class, immigrant in all one's intersectional complexity inflicts psychic damage and curtails developing scholars' creative potential. During my PhD training, I found a lab that worked differently—the Carlson Lab at UC Berkeley, which was majority queer, half people of color, and genders all along the spectrum. There I experienced a feeling of tremendous relief, an unfurling of my shoulders and loosening of the chest. In the field with these queers and accomplices, I could open my heart to our common love for animals, for precision in observation, and for puzzling through relations and response—without constantly anticipating a racist aside or a misogynist joke. All that cognitive energy was then available to explore the ethics that drove our engagements with science and with the species and places we study. In talking about how those commitments and modes of study are shaped by our experiences, our different cultural and embodied knowledges became visible, and they deepened through sustained collective thought and feeling.

4

Unchartable Grief

Scientists Grapple with Extinction Politics

> What would it mean to consider the land itself as a site of an agentive fungibility that has been conscripted into the proprietary spatialities of colonial possessiveness and constrained into geographies of exploitation that no longer serve the relationalities of presence and care that have for so long been its domain as a common for all?
>
> —JODI BYRD ET AL., "Predatory Value: Economies of Dispossession and Disturbed Relationalities"

> The subject of reason relies on objects that remain external to it for its own self-constitution. Even the most abstract of thoughts results from the movement of matter on matter. Individuation transpires through the vulnerable process of object relations. The relational, rather than autonomous, development of the individual in which all external to the self serves primarily to further its own development is a notion suited to colonial economies, in which value is extracted through the subjugation of lives, lands, and resources figured at geographic and/or temporal remove.
>
> —KYLA SCHULLER, *The Biopolitics of Feeling: Race, Sex, and Science in the Nineteenth Century*

To incite field ecologists to take up a queer politics that unsettles the proprietary spatialities of colonial possessiveness, I theorize a relational research methodology that arises in a nexus of love and grief in ecological fieldwork: the death and impending extinction of beloved species. This politics allies with Native land struggles as it flows from queer radical traditions. The Stonewall Rebellion, ACT UP street protest, vigils for murdered queer and trans people (overwhelmingly Black, Indigenous, and immigrant), pro-Palestine organizing, and uprisings for Black lives—queers make kin in the streets and practice care in the face of state violence, channeling love but also rage and grief. Queer relational methodologies can be underflows to disciplinary practices in environmental science, eroding norms and mentorship structures that long excluded lesbian, gay, bi, and queer researchers from intellectual belonging. Using this approach, queer and trans scientists can bring our full selves to scientific practice; others cast out of straight time's rhythms can also take up these tactics. Feminists, environmental justice researchers, and anyone whose scientific work transgresses the university's boundary do queer ecology, not by identifying as queer (read: deviant in sexuality) but by doing queer collective work to protect lives and relationalities that have been devalued by heteropatriarchal capitalist logics. For non-Native queers and outcasts, especially those cut off from families and landscapes of origin, collective care for the land in rural collectives often engenders powerful attachments to community, while perpetuating settler-colonial erasures and white supremacy. Byrd's naming of the land itself as a site of agentive fungibility as a locus of kinship and relationality for Indigenous peoples suggests a path for queer politics committed to unsettling land, while extending queer theories of kinship to encompass other-than-humans.

To develop this approach, I move between modes of writing: a transcript of an extemporaneous science talk, the formal humanities paper, and an interactive, reflective mode with echoes in feminist, queer, and performance studies. My site is, again, young salmon facing extinction, but they could just as easily be other species that face deliberate or inadvertent death driven by settler-colonial logics of domination. I perform a queer reading of my own plots and graphs showing that juvenile steelhead coho and steelhead in Salmon Creek tributaries died in greater numbers during the driest years of the 2011–15 California drought. I think with ACT UP scholars to float a hypothesis: by grounding public activism in our love and care for our study species—a form of practical kinship—we scientists can enhance our capacity to feel, to grieve, and to take action from that grief.[1]

Representing the Field [Transcript]

This is a field site.

It's a small stream where I've spent countless hours observing water, sediment, and fishes.

Do you have a field site? Imagine you are there. Close your eyes if you like.

Is there a breeze?

Do you have prosthetics that let you see farther, or sense things not seen?

What are you listening for?

Is there water touching your skin?

Who is with you?

Are they dying?

If so, are they dying too soon?

How does that make you feel?

Now open your eyes.

When I present about my field site, I often begin with a slide showing a panoramic photographic collage of a small pool in a dry streambed. In ecology, we are trained to find field sites that reflect larger phenomena, which are usually abstracted from places made up of culturally situated communities and their more-than-human relations. In a small gesture of refusal of scientific norms, I evoke those place abstractions with the phrase "salmon live or die in little streams." By avoiding Latinate words with scientistic connotations, I try to talk about the fishes' life chances in the language of kinship and relation. I'm not sure whether the ecologists and hydrologists in my audiences notice this gesture.

The photographs on the next slides show me working in the field with other scientists and local residents. I describe my observations of this site, and the analyses of those observations, in technical language. These mostly Latinate terms describe with precision a set of standardized methods. The language obscures the sensory: the grit of icy sand and gravel through my waders as I kneel to dip the dissolved oxygen probe in the water, the crick behind my left shoulder blade as I stoop over with a two-pole seine, the taste of anaerobic muck in my snorkel.

Next, I show results of statistical analyses and a map I made from data that I collected together with local citizen scientists and residents to connect the salmon deaths to the longer time that the fishes spent in isolated pools during the drought years (figure 4.1). I include a photograph of my research collaborators collecting this data, but I never, in scientific talks, have time to share our field conversations about the fishes, about local politics around groundwater and pumping out of the stream, or about my collaborators' joy, one day out of the year, at seeing "their" fishes flitting about the small pools. Instead, I use the words *drought, mortality*, and *persistence* to transform our observations into abstracted points, lines, and numbers on the maps.

Natural historians could, early in the last century, publish papers describing a species or change in population, but now ecologists must propose a mechanism for this change. The change I am interested in is the number of salmon fry that survive summer in intermittent streams or isolated pools, and what story these numbers tell about whether salmon will keep returning to these streams or go extinct. As I move through the slides, salmon become more and more abstracted, from individual fish to an aggregate outcome, pool-scale persistence, or the question of whether any fish survived the summer in a given pool. Now, one slide proposes low dissolved oxygen as the mechanism of salmon death over summer and replaces data on actual salmon with a threshold determined by other studies. Other researchers determined a "sublethal limit" and "lethal limit" by placing fish in low-oxygen water in a lab and observing how they responded. My slide shows plots of dissolved oxygen levels in three pools: one intermittent pool where water keeps trickling, one isolated pool that has no flow for two months, and one isolated pool that has a tiny spring trickling into it from above. In each plot, the dissolved oxygen drops past the sublethal threshold, when salmon stop being able to move around to feed and become more vulnerable to disease. In the isolated pool, the dissolved oxygen level drops below the lethal threshold in late August and stays there until November; fish suffocate. In these conditions, I tell the audience, no salmon did survive. In the other two plots, oxygen levels stay in the sublethal zone, barely; I tell the audience that some fish did survive in these pools, but they were stressed, weak, and too small and starved to have much of a chance once they washed downstream to the estuary.

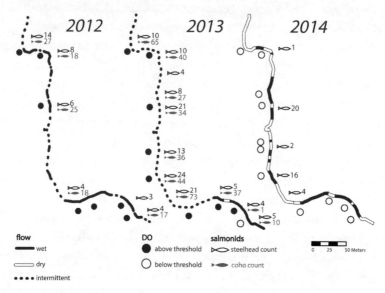

Figure 4.1. Map showing wet, dry, and intermittent stream reaches, oxygen levels, and fish presence on lower Fay Creek, 2012–14. Figure by Cleo Wölfle Hazard.

Each of these three pools is a real place, which I know intimately by the depth of water on my waders, the seep of water through holes in the knees of my wetsuit. The stream in fall is a collection of pools, some with sandy bottoms, others with rocky hidey-holes and shallows where coho and steelhead congregate. Between these small pools are expanses of dry streambed, made up of large cobbles wedged into gravel. These bedforms are echoes of floods, left behind when raging torrents unleashed by the winter subsided. Because of these flashy, powerful flows, the stream shifts form over years. Many pools move from place to place as sediments fill in some places and are scoured out of others, while others maintain their shape because of large boulders or sharp bends in the stream channel's clay walls. I wrote pages of notes on natural history in my field notebook and sketched and labeled dozens of maps.

Here's what made it through revisions into the published paper:

> The upper and lower Fay Creek study reaches flow through a broad, low-slope alluvial valley, where large pools form at occasional clay or bedrock outcrops, sharp bends, and large woody debris jams. The lower

Tannery Creek reach also flows through an alluvial valley but is more confined and steeper than Fay Creek, and as a result has many more small, shallow pools. In contrast, the upper Tannery Creek reach flows through a steep canyon in a channel that has incised into the confining clay and bedrock layer, but large redwood logs and root wads create some large pools within the shallow veneer of alluvium. Reaches with deep alluvium become intermittent before bedrock reaches, but also contain deep pools and complex habitat that characterize preferred habitat for juvenile coho salmon.

Study reaches ranged in flow status both within and among years, with upper and lower Tannery Creek reaches maintaining continuous flow in 2012 and 2013, the lower Fay Creek reach becoming moderately intermittent in all years, the upper Fay Creek reach becoming extremely intermittent in all years, and no reaches maintaining continuous flow through the extraordinarily dry summer of 2014. During the driest conditions, we encountered subreaches that maintained adequate hyporheic and/or spring flow to sustain juvenile salmonids.[2]

Our results also point to the importance of spring-fed tributaries that tap shallow groundwater sources for sustaining salmonid fishes through drought, suggesting that spring-fed tributaries can play the same role as hyporheic flow through isolated pools in elevating oxygen above lethal limits. Spring-fed tributaries delivered relatively oxygenated waters to a subset of our study pools. Despite often flowing at a rate under 1 liter per second, these inputs maintained pool DO [dissolved oxygen] levels >7 ppm through the dry season in the wetter years and less than 3 parts per million (sublethal level) in the extraordinarily dry year (2014), despite more than 100 days of disconnection. At a regional scale, spring-fed tributaries, critical for maintaining sufficiently high DO for salmonids, occupy a relatively small proportion of overall stream length. Our results add to research on other taxa that suggests spring-fed refugia within intermittent streams can support high densities of imperiled organisms. Our results also suggest that, to function as a sanctuary reach through periods of drought, a spring-fed stream must also feature pools that are large enough to maintain pockets of adequate DO through periods of drydown, or that have continuous hyporheic flow through the pools to prevent source depletion through respiration and stimulate oxygen reaeration and mixing through the pool. More work is needed to understand micro-scale variation in DO within pools and how it may be controlled through hyporheic exchange, air–water gas exchange, and thermal convection. This study also underscores the imperative of obtaining a more detailed understanding of the hydrology of intermittent stream watersheds, including flow paths from diverse aquifer sources with distinct contributions to in-stream water quality and habitat.[3]

I made many maps to try to show these relationships between where in the stream a pool was located, how much groundwater flowed into that pool, and how salmonids responded to those conditions. The one I settled on shows the stream as a wiggly line, red if it's dry, blue if it's flowing, and a dashed line if it's intermittent. I represented the steelhead and coho as colored icons and the pools as circles (red for lethal oxygen levels, blue for survivable oxygen levels). Figure 4.1 shows a simplified version of this map, for one stream reach, in September in three increasingly dry years.

Returning to the talk, I next show that oxygen alone did not explain fish death. I show the survival rates in a different way for all of the pools I measured over three years. Each black, white, or gray dot represents a pool, either one where any fish survived or one where they all died. Some died quickly in a heron's beak. Others starved slowly or ran out of oxygen. In most pools, a few fish persisted until the fall rains brought flow and oxygen; in a few pools, fed by trickling springs, more than one hundred survived. The left panel in figure 4.2 shows this data; the right depicts how different habitat conditions explain death and life or, as ecologists say, persistence. I remember many of those fish, not as individuals, but as lively presences I'd see eight or nine times over the summer. They grew bigger, then sometimes, suddenly, they were not there, or their whole pool dried up. Twice a year I'd count them while swimming or belly crawling through the pool. To count salmon this way, you need to first estimate roughly how many there are as they swirl and dart in the shallows, then track them one by one (by species, but also by three different size classes) as they flit past between your body and the bank, and then *not* count them again if they dart back upstream.

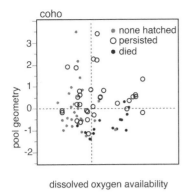 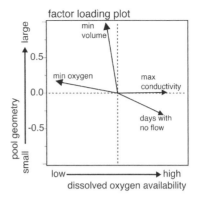

Figure 4.2. A representation of how habitat factors influenced whether salmonids survived or died in Salmon Creek pools. The left plot shows pools, represented by dots. The right plot shows the habitat factors I measured. This visualization is called a principal component analysis. Figure by Cleo Wölfle Hazard.

I close this portion of the talk by asking the audience to close their eyes again, and I ask

> What do you feel, looking at these charts?
>   Did you think of your field site?
>   Of your own study organism?
>   Of your beloved species and its vulnerability as a population and mode of life?

On Feeling Too Much or Not Enough

For the "we" who are field ecologists, the problem of feeling too much or not enough involves (at least) two questions: "How do we not feel the death too much?" and "What is a better way to not be burnt out by our work?"[4] As ecologists, can we study extinction and contamination without shutting down our feelings of grief and loss, instead tapping into emotions as a source for social connection and political action? I discussed the first question in chapter 2. Ecologists are trained to excise mentions of grief and sadness at premature death and suffering from our papers and conference presentations.[5] The world, our focal species and its coinhabitants, speaks to us through the data. Often the data we collect shows whether an organism lives, dies, grows, reproduces, migrates, or thrives given certain conditions. Many ecologists begin to self-censor expressions of this grief in the field, at least when they're out there among colleagues. Emotion and affective connection are framed as unscientific because they are unobjective, a possible source of bias in statistical analysis or interpretation. Yet in making space for affect and feeling to come back into field sciences, it is crucial to understand care as noninnocent and a primary tactic of racializing power. Kyla Schuller notes regarding nineteenth-century biology that the idea of feeling and affect—*sentiment* and *impressibility* in the terms of the day—became sutured to whiteness through the work of the American School of Ethnology. Early naturalists attributed their observational and analytical powers to a finely honed, but not excessive, sensibility. In reviving care and concern in feminist new materialisms, and in calling for ecologists to not suppress love, grief, worry, and other affects associated with care, I keep Schuller's caution in mind.

We ecologists and field biologists are trained to be curious about what the data show, whether or not those results match our expectations or desires. When the data show loss—of a special place, of an individual creature to poisoning, starvation, or suffocation, or of an entire species to habitat loss or a changing climate—this can be a good scientific result. "Good" means that our study design was robust enough to detect a change and our models suggested a mechanism for that change. If we feel a thrill in that discovery, in that elegant meshing of theory and data, then this emotion is welcome in a conference talk or even, in a measured way, in a research paper. But we ecologists rarely share the loss of a beloved study species with our collaborators or in public by lamenting and grieving a stonefly's disappearance from a desert spring or the absence of ten thousand rasping monarch butterfly wings from a steep pine mountain or a yellow warbler's song in spring. In our repeated self-silencing, I worry that we lose some capacity to feel, to grieve, and to take political action from that grief.

Collective acts and rituals of grieving are always about love for the living and a deepening of bonds through shared experience. During the height of the AIDS crisis in the US, queers and other outcasts turned grief at the mounting deaths of their chosen kin toward public performances of mourning and rage. David Wojnarowicz's *Close to the Knives*, Eve Kosofsky Sedgwick's "White Glasses," Leo Bersani's "Is the Rectum a Grave?"

Deborah Gould's *Moving Politics*, Douglas Crimp's "Mourning and Militancy"—works reflecting on this time all convey a sense of collective breathing space that this acting-together-in-the-face-of-extinction created. What was unbearable as individual grief became a cathartic, collective rage that spurred political action for medical research, for antidiscrimination legislation, for social support for people living with AIDS, but above all for visibility and an end to homophobic erasure of the epidemic. In mourning beloved dead in public, the thousands of friends and lovers who came into the streets as ACT UP proclaimed that these dead had lived lives worth grieving and demanded an end to policies that made HIV-infected bodies killable by neglect.

Although I was a child during the height of the AIDS crisis in the US, I heard queer and trans elders relive these experiences later, in the streets at vigils for trans people killed by friends, strangers, or the police. I wonder, as I think back on my own encounters with dead and dying study organisms, whether such public mourning for interspecies relations could bring new fire to environmental activism and perhaps replace affectless terms like *species loss* and *biodiversity* with *death of beloved kin*. Mindful of Michelle Murphy's caution against "the conflation of care with affection, happiness, attachment, and positive feeling as political goods," I focus on grief as a negative affect that is sometimes conjoined to love and other times infused with rage, frustration, and despair.[6] Murphy's "vexation of care" draws on Sara Ahmed's attention to alienation from normalizing strictures, to contentment that arises from being aligned with dominant ways of belonging.[7] Like Ahmed, I am most interested in affects that arise in moments of alienation—from settler discourses of environmental stewardship that sacrifice riverine ecocultural relations for extraction that props up settler collective continuance. Alienation from settler normativity arises in moments of unease at settler water management's collateral damage. Murphy asks, "What, then, is the work of discomfort, unease, and trouble in matters of care?"[8]

Practical Kinship and the Politics of Shared Grief

If salmon could be kin—if ecologists could, through caring for and closely observing and working to improve the life chances for a given organism, come to feel a kind of queer kinship with the fish—what would that mean, and how would it happen? Let's first think with Elizabeth Freeman about "the possibility that queer bodies make something that might at least theoretically extend and endure."[9] Freeman is open to the idea that kinship may not be desirable for radical queer theory and politics, which reject assimilation, but she nonetheless considers queer kinship's power to denaturalize heterosexist norms. Riffing on Judith Butler's *Bodies That Matter*, Freeman argues that "(1) a culture's repetition of particular practices actually produces what seem to be the material facts that supposedly ground those practices in the first place, and (2) when those repetitions are governed by a norm, other possibilities are literally unthinkable and impossible."[10]

To shift away from an idea of kinship rooted in procreation and lineal descent is, for Freeman, to focus on kinship as a "resolutely corporeal" practice, "a set of representational and practical strategies for accommodating all the possible ways one human being's body can be vulnerable and hence dependent upon that of another, and for mobilizing all the possible resources one body has for taking care of another."[11] With this shift, queer modes of relation become intelligible as kinship. To extend this argument beyond relations with human beings, I understand scientists' practices of paying attention and taking care as strategies that mobilize all the resources one (human) body has to give to a (nonhuman) body. By caring for a nonhuman other, scientists come to care about that other and perhaps become vulnerable to grieving for and loving it. We scientists depend upon its continuance as a species not only for our livelihoods but also to renew a sense of purpose in our work to help its life-chances.

The kinship that Freeman is interested in for queers does not reproduce the genetic family, but rather "reproduces the cultural force insofar as it also recreates and recharges bodies toward ends other than labor, such as play, love, and even violence. So does queer life, though perhaps in ways for which we do not fully have a vocabulary."[12] Nonhuman beings excel at what we humans name play, love, and violence, and we theorize from these actions to our own ends. I did this when I theorized beavers' refusal to be co-opted into modes of productive labor via the archetype of the "ecosystem engineer." Perhaps if queer ecologists theorize more-than-human kinship, we can replace such mechanistic models of ecosystem function, rooted in control and extraction, with queerly anarchistic models of how to live together.

Freeman goes on to formulate a kinship rooted in desire for rich and sustained connections. Judging from my conversations with ecologists in the field and after giving versions of this chapter as a talk, I think many ecologists' longing and desire to belong in a web of more-than-human relations powerfully motivates their scientific and conservation work. Freeman's formulation of kinship as belonging (both the true etymology of *gelang*, "alongside of" or "at hand," and the false one of "being long" in time) is important to my argument, so I quote a key passage at length:

> "Queer belonging," I would therefore argue, names more than the longing to *be*, and be *connected*, as in being "at hand." It also names the longing to "be *long*," to endure in corporeal form over time, beyond procreation. Though I offer a false etymology here, "belonging" contains the verb "to long," from the Middle Dutch *langen*, to be or seem long; "to 'think long', desire; to extend, hold out, offer." To want to belong, let us say, is to long to be bigger not only spatially, but also temporally, to "hold out" a hand across time and touch the dead or those not born yet, to offer oneself beyond one's own time. Longing to belong, being long: these things encompass not only the desire to impossibly extend our individual existence or to preserve relationships that will invariably end, but also to have something queer exceed its own time, even to imagine that excess *as* queer in ways that getting married or having children might not be.[13]

Freeman here is writing about queer families, of ways we make kin that include but also exceed experiences of gestation and child rearing, which punctuate ongoing sexual and political bonds. For scientists who study climate adaptation over centuries, who imagine populations continuing and expanding even in the face of increased pollution and human extraction of water from rivers, this being long extends across space and beyond individual bodies of plants, animals, fungi, and microbes to populations and evolutionary potentials. Anna Tsing describes matsutake mushroom scientists who volunteer to restore red pine forests;[14] these scientists imagine temporal relation longer than the queer ancestry Freeman invokes through drag performance. Their recognition of symbiosis that coevolves with human cultivation reshapes science practices to a more relational mode. The idea that scientists could study places they also care for is implicit in Donna Haraway's image of companion species "becoming with" one another in evolutionary and everyday relations and explicit in Tsing's rejection of plantation science, where "love does not flow between expert and object."[15] This rejection of plantation science and embrace of care makes space for queer and otherwise marginalized scientists to study our beloved streams and creatures. In rejecting plantation science, we embed care work for one into field protocols and other institutional structures of science. Safety plans to protect against race-based or sexual harassment, field practices that reduce bodily injury, health care, leave policies, and collective care practices can make it possible for BIPOC, queer, trans, and disabled scientists to do field-based work.

To recapitulate the argument so far, field scientists, whether queer or not, may form queer kinship relations through field practices that perform caring and vulnerability and thus engender practical kinship. Recognizing these relations as kinship could help field ecologists communicate their data and field relations differently, in a public politics that demands collective care for their beloved kin. This politics would mourn premature death—human and nonhuman alike—from resource extraction, pollution, and climate change. Many environmental scientists already join with local communities or work in the trenches with activists. Perhaps queer ecological politics of kinship will resonate with these people, queer and otherwise—and inspire them to deploy their research in environmental struggles.

On Salmon Creek, Talking to Fish Who May or May Not Die Young

Let us now consider how queer relational methodologies—ways of making kin—could enable people to do science differently. Staying with Freeman, how can practices of queer habitus (from Bourdieu, "the durably installed generative principle of regulated improvisations") move between species?[16] Commenting on how habitus communicates from body to body, Freeman writes:

> *Habitus* is, in Bourdieu's words, a kind of "plastic art . . . a *mimesis*, a sort of symbolic gymnastics": the term names the way that bodies become similar and hence attuned to one another outside of a theory

or even of language itself. As a practice, *habitus* "communicates, so to speak, from body to body; i.e. on the hither side of words or concepts." Especially given the emphasis on bodies here, we might think of this "communication" less as a verbal transaction and more as a kind of contagion, in the sense that diseases are communicated.[17]

Does contagion pass between human salmon ecologists and the fish they swim among? For Freeman, "the point is that 'kinship' itself is actually 'the field of relationships constantly reused and thus reactivated for future use.' Kinship consists of relationships renewed, and their very renewal is what is relational at all."[18] This may happen to an ecologist in the field at a long-term study site, conducting the same measurements for the hundredth or thousandth time, to track dynamic interactions among organisms, currents, climate, and disturbance. Or it may happen when one thinks of the individual fish one measures as part of a population, then associates these fish with others at a different site and imagines their relation through past or future evolution. If we think of affect as able to act at a distance, then we start to understand how feelings of grief, rage, frustration, or joy, which arise as field scientists track beloved species in their unpredictable movement of a population to extinction, can engender radical politics of refusal. These politics could channel scientists' alienation from the normative systems of research and governance in which they usually operate, inspiring them to march in the streets or place their bodies in the way of pipelines, herbicides, or bulldozers.

At my own long-term study site in Salmon Creek in central California, where I spent seven field seasons studying juvenile salmon survival, through a deepening drought followed by spectacular floods, I became intensely curious about how and why coho salmon and steelhead choose the pool where they spend their first long, dry summer. In early spring, the young fish can swim freely upstream and downstream, but then they are stuck in whichever place they have chosen once the shallows dry into long, dry stretches of gravel. When salmon disappeared from one week to the next, I looked for raccoon or bird tracks, for any dead bodies, and for clues in the dissolved oxygen readings as to whether they had suffocated or become prey.

Sometimes, when I was alone, I would talk to the small fishes and say I was sorry they were starving, were struggling to breathe, were frantic, hiding under leaves from the big loud mammal. But I didn't rescue them and move them to a different pool, where they might survive. I didn't move them, in part, because other fish already inhabited those deeper pools. Adding fish might mean that none survived. But also, I wanted to know why fish survived in some pools—I wanted to understand processes, causal mechanisms, the dynamic links between flow, habitat, and the species' evolved adaptations to that environment, now changed by drought and human water extraction. I am struck, on writing this, by how, when one speaks the language of ecology, it's hard to escape industrial and mechanistic language and hard to think about how to introduce the language of emotions. I worry that such language short-circuits contagion by limiting the modes of relation that scientists and nonscientists can cultivate in work toward shared goals.

Early scientists like Darwin and Fabre mixed expressions of wonder and delight into their descriptions of species morphology and behavior.[19] Writing of Darwin's descriptions of bees and flowers, Carla Hustak and Natasha Myers coined the term *involutionary momentum* to describe the relational, even affective dimensions of pollination as an "'ecology of practices' among plants, insects, and the scientists who draw their intricate relations into view."[20]

> [Darwin's] work on orchids wrapped him up in the daily rhythms and mundane encounters of mingling species whose strange and unfamiliar intimacies challenged assumptions about both bodily and species boundaries. Indeed, as Darwin trained his attention on the intimate encounters between orchids and their insect pollinators, his functionalist accounts of adaptation were sometimes muted by stories of affinities, attractions, and intimacies. Darwin's own experimental practice takes the form of an *ethology* sensitive to plant and insect practices. As we track his multisensory experimental techniques, we find him leaning into the event of pollination to mimic both insects and orchids as he attempts to generate experimental proof of the mechanisms of fertilization. Along with the insects, Darwin was caught in the orbit of these alluring plants. . . . [H]is inquisitive, multisensory experimental practice got him *affectively entangled* in the event of fertilization. When we amplify Darwin's modes of attention and involvement in the daily rhythms of life among insects and orchids, we find in his account the

nascent contours of an *affective ecology* forming the grounds for a science of interspecies relations.[21]

Their argument hinges on Darwin's body, which takes on plants' shapes and mimics bees' movements; the *momentum* of involutionary momentum is "less in the sense supplied by Newtonian physics and more in the sense of what dancers may feel as they lean into and follow through on a movement; that is, as 'an impetus' and as the continuing vigour resulting from an initial effort or expenditure of energy.'"[22]

It is this account of affect traveling between bodies (the bee's hairy body delicately brushing the orchid's pedicel, which responds by forcefully ejecting a grain of pollen) that resonates with Freeman's queer reading of Bourdieu, which I present shortly. Hustak and Myers focus on practices of science and propose a lively relationality to inform scientific practices with plants, in order to challenge "the status quo of ecological responsibility."[23] I propose to extend this relationality into the realm of politics, to consider what political interventions ecologists could make on behalf of species they know only indirectly, following a model by which queers imagine ancestry and kinship across space and time. Freeman writes,

> Official kinship, Bourdieu argues, is relatively inert in everyday life. Practical kinship, on the other hand, is ubiquitous. . . . What interests me about Bourdieu's model is that he grounds practical kinship in a specific form of "shared substance" between bodies, the concept of *habitus* that he borrows from Marcel Mauss. This term roughly translates as a learned bodily disposition, stance, or schema. Moreover . . . the importance of temporality to *habitus* provides a way of thinking about queer belongings in a temporal as well as a spatial sense: as modes of duration not only *for* otherwise mortal bodies, but *between* bodies otherwise separated in time.[24]

Scientific mores no longer permit such diversions as Darwin and Fabre made into intensifications of feeling in the field with insects and earthworms and flowers. At my Salmon Creek field sites, grief at salmon death took hold of my body and my breath and reverberated with the chill autumn melancholy of the dry streambed littered with fallen leaves.

Later, back in the lab, I typed numbers from data sheets into spreadsheets and graphed those declines in numbers of salmon by writing statistical analysis code. I would sometimes recall that field affect in an echo of bodily posture, slumped shoulders, pain in between the shoulder blades, eyes losing focus and seeing again the black water in my mind's eye. Did the hours of computer work create the back pain that recalled other back pain? Or did that pain come from stooping again and again under willow thickets while working a two-pole seine toward a point bar to scoop up young coho and steelhead fry? Or did the grief, remembered or reimagined, call up pain that then evoked the field? These questions of somatics and the body seem, in hindsight, crucial to any scientific insights that those figures and statistical models revealed. Surely these feelings influenced which of several possible statistical analyses I performed or my interpretations of their resulting probabilities. Yet I didn't even try to write those feelings into my data analysis, even in the more speculative discussion section, having been disciplined by professors and coauthors out of undermining my authority as an objective scientist. Feelings, intuitions, or political commitments might contaminate data analysis. This emotional disconnect, however generative of ecological models, limits the work that ecology can do in the world, because ecologists hide the feelings that could draw others into their dreams for future ecologies. False objectivity constrains our ability to respond.

On Staying Open to Suffering: Ethics in Field Methods

Field conversations with my lab-mates shaped my thinking on ecologists' affects toward dying field organisms. Two years after Suzanne Kelson and I counted salmon on Salmon Creek in the depths of the drought, I drove up to her Eel River field site a few hours further north, to help her survey steelhead fry on a tributary. I was excited to see this protected stream deep within the University of California Angelo Reserve, to practice electrofishing, the "gold standard" technique for surveying fish in streams and rivers, and to meet some nice young fish. With a crew of five undergraduates, we backpacked the forty-pound electro-fishing current generator, extra batteries, and assorted buckets, measuring devices, and fish tagging equipment down a trail, then up a steep, boulder-bed stream thick with lush growth. Water flowed quick and clear in the shallow, rocky pools of the headwater stream and wide and sometimes deep in the larger tributary—a far cry from the putrid anoxic pools we had surveyed at Salmon Creek. Suzanne wore the electrofishing backpack, trailing a metal lead in the water. To catch fish, she passed a long metal wand over the streambed and jabbed it up under the bank. While the electro-fisher ran, it beeped and hummed. Three or four undergraduate assistants followed with dipnets, looking for a white flash when a stunned fish flipped on its side, then darted in to scoop it up. When a fish "flipped," it stilled only for a second or two. If the netters missed, the shocker shocked again—everyone hoped that someone would land the fish in the net before its heart stopped permanently.

We put the netted fish into perforated buckets in the cool stream, then Suzanne and a student assistant worked them up. They anesthetized them in a bucket with two Alka-Seltzer tabs dissolved in it, producing carbon dioxide that quieted their darting. The student weighed and measured them, and then Suzanne photographed them and scanned them with a microchip reader, to see if the fish was previously caught and tagged. If not, she clipped a piece of the tail for genetic analysis, made a small incision on its belly, inserted a rice-sized microchip, and then stroked the incision with her forefinger to reseal its protective slime. Back in the aerated recovery bucket, most fish started flopping in my hand or swimming quickly once I opened my hand to release them. Over the course of the three days, three or four (of around one hundred) didn't recover. These "morts" fall well within the limits of Suzanne's scientific-collecting permit, but the whole crew was sorry at their deaths.

By instinct honed by years of counting salmon at my Salmon Creek field sites, I counted the fish as we approached the pools, sometimes not realizing I was counting, but holding the tally in my head nonetheless. Even for a queer and trans scientist thinking about the emotional content of fieldwork, I had to practice *not* immediately counting and measuring and instead let feeling sit in my body and come into conversation and field notes. The steelhead fry were easy to spot in the clear water (if one knew what to look for), though a few hid under the banks. I noticed that, especially in very bouldery pools, we netted fewer than my count. When I mentioned this to Suzanne, she said she's noticed the same thing. She regretted the deaths and questioned the "old guard" logic that shocking doesn't harm the fish or stunt their growth. "I think we could get better data by snorkeling, estimating length visually, and then fly-fishing to get fin clips for genetic analysis," she said. If she were to start the study anew, she wouldn't use electroshocking as her main method. An undergraduate researcher told us that she didn't like how other students collected animals, saying, "Sometimes they handle them for fun, or to make a video, not for data. I don't think that's right." Over the course of two days in the field, as we revisited the topic between sampling efforts, Suzanne decided to do an extra day of snorkel surveys, then compare the counts to the electro-fishing method, and eventually write a paper.

Like medical doctors, field biologists and ecologists are trained *not* to feel individual grief at killing bugs or at letting fish die. This training provides an emotional buffer for scientists who, deliberately or accidentally, kill or harm animals during research. Formal ethics training teaches that animals should not suffer pain or be killed unnecessarily. We fish ecologists transmit the nuances of these ethics informally when we explain a field method, watch fish respond to electroshock or sedation, and feel their still or flopping bodies in a water-slick hand. We debate the most ethical ways to capture and tag. Some feel strongly that electrofishing kills fewer fish, while others, like my own field mentor Michael Fawcett, feel that net seining causes less injury and death. A scientist who collects and kills aquatic insects told me he felt bad when he watched them squirm as he poured ethanol over them. He thought they were struggling to escape and live, even if they lacked pain receptors or memory pathways.

Should research ethics change when research is trying to stave off extinction? Does extinction have its own affects, and if so, does ecological training enhance or diminish our capacity to grieve for populations and species under threat? As humans, can we repress love and grief and fear at extinction without closing off emotional ways of learning and seeing differently? For me, the process of reckoning with the ethics of different kinds of death enhances my capacity to grieve, but I'm still struggling to find a way for that grief to bridge science and radical politics for ecocultural survival. Scientific practices of minimizing the emotional impact of other-than-human premature death prop up settler logics and extractive projects. Scientific norms of "objective" writing—of imposing emotional distance when we write of death and suffering—frustrate my reckoning with grief as a source of queer ecological politics of mourning.

Thinking Queer Relational Methodologies as a Radical Tradition and Politics of Solidarity

The only way I have been able to keep working in the field as a scientist is to do that work with people who work not just to restore ecological processes but also to repair cultural relations along rivers. In earlier chapters and interludes, I've introduced you to many of them, but I want to return now to the people I worked most closely with while writing this book: members of the Salmon Creek Watershed Council in Sonoma County, California. Like those of many such volunteer organizations in the US, its members are mostly white, many of them are women, and several are retired. The people I spent most time with included a retired professor, a shepherd and wool artisan, a retired computer programmer turned amateur cheese maker, a classical flautist, a retired tradeswoman, and an environmental consultant trained as a geomorphologist. As a beginning graduate student who had worked for a decade in river restoration and household water conservation, I wanted to work here because of the council's focus on restoring streamflow. At that time, in 2010, few stream restoration projects studied how rainwater harvesting and groundwater recharge for residential use would affect streamflow and aquatic habitat. For two decades, this group and their loose network of neighbors has been reshaping their own water use and local governance practices to give salmon a better chance at surviving the next drought. My seven-year collaboration with them began as a social study of household water supply. Eventually, we added on an ecological study of streamflow and salmon survival through summer during deepening drought, then codesigned projects with a collective of local residents and agency scientists.

What I want to think about here, in light of the above reading of practical kinship, is whether my intervention there was a queer one. Did my thinking about salmon kinship and sharing experiences of grief in the field make a queer movement toward more-than-human relations that enacted a different kind of politics? My Salmon Creek collaborators built social ties at the local bar, by sharing water stories of our own failing springs and frozen piles, and at annual parties and festivals. The council met every month, and we all came into relation with salmon and their habitats together, observing, mapping, and measuring the stream. The council members had started to untangle the many factors contributing to salmon extinction. They were monitoring water quality and thinking about taking other action. But, by their own reports, my intervention galvanized them. I pointed them toward methods that they found meaningful and useful. These methods deepened their understanding of ecological dynamics of streamflow and salmon survival and brought many of them face-to-face with local salmon for the first time. Meeting the vulnerable fish in streams connected to their own wells made some residents use less of that well water. Perhaps the repetition of this performance of care through data collection and visualization—in the field with me and monitoring rainfall in their rain gauges—made them see and feel with salmon differently. In early interviews, several Watershed Council members described salmon in symbolic language—stand-ins for wildness and the web of life—or as a commodity—reason for a sportfishing trip, or a (vanishing) source of livelihood for family members who fish commercially. As they collected and processed data on patterns of wet and dry reaches in streams, these residents-turned-community-scientists came to see salmon populations in a way that Western ecologists and Indigenous scientists do—as embodying specific life histories, as keystone species that materially connect the oceans with mountain forests when they become food for terrestrial predators.

The Watershed Council members who came out into the field with me were entranced and charmed by fishy vitality, as was I. Year after year, we would meet up in the field; they helped me at first, and then, as they took over the wet-dry mapping effort, my students and I helped them. In the field, when we saw fish crowded into a shrinking pool, they would often ask, "Why can't we save the fish that will die in this tiny pool and move them to another pool or another stream?" Following evolutionary logic, I would answer, "Because if there were other fish in the new pool, increased competition for food might lead them all to starve, and moving them might make their descendants less fit to survive low flows." Then, sometimes, they would ask if I was sad that the fish would die. I was often sad, out in the field, but sometimes I would try to explain the different nuances of sadness—what Deborah Bird Rose has articulated so eloquently as the double death of extinction,[25] which feels different, and matters differently, from the expected death after spawning, or the predicted death of most of the fry even under the best environmental conditions. To me, thinking as an ecologist, the two did feel different, but in the field, statistical matters receded. These conversations with my collaborators helped remind me to also feel and grieve the individual deaths that came too soon.

I remember one time, in the field with Erna and Dave. Erna was devastated to find so few steelhead; her excitement at being on the creek turned subdued; her shoulders slumped. This was in 2014, the depths of the drought, and I had done enough data analysis to know that the numbers of fish we were counting trended toward extinction. I said so, then realized from their dejection and disappointment that I had just performed scientists' doom and gloom to a tee—instead of responding with my own feelings, I privileged data and mechanistic response as I had in my science paper. I recognized my pre-scientist self's response in their emotionality and grieved that self's fierce sense of right and wrong. In witnessing Erna, Dave, and other local volunteers' dedication and care for the fish, I built relations that became an affective underflow that helped me resist straight science's norms. Part of queer field kinship is letting ourselves—as scientists or fishers or swimmers or amateur data collectors—be drawn into relation through sharing grief at fish death and then joy at fish life, like fabulous queers at a public wake. Joy and grief together make an affective circuit as intensification, which extends to and through the fish and water.

In my elaboration of queer field affects as collective and generative of oppositional politics, I join Murphy in centering the meaning of care as "troubled, worried, sorrowed, uneasy, and unsettled."[26] This core sense of the care arises in moments of collective harm, in the action of washing pepper spray out of a stranger's eyes at a street protest, in witnessing deaths by state violence, or in reclaiming vacant land for housing and gardens. Practical kinship beyond the human has at its target biopower, which, as Schuller notes, "has consolidated at the site of the body's capacity to perceive, feel, emote, and express and thereby modulates our very sense of collectivity, of communion, and of companionship. Race, sex, and sexuality coalesce as the products of our sensory stimulations, emotional entanglements, and affective interactions."[27] Responding to calls from Black and Indigenous queers and feminist leaders of contemporary justice struggles, Schuller argues that "to restructure the political, we need to reconceive the very grounds of the organic and the social."[28] Returning to the epigraphs, queer feminist field practices must critically interrogate concepts of relationality and the commons that have, on the one hand, enabled settler-colonial dispossession and white supremacy yet also retain valences of resistance to settler enclosure and modes of domination within Indigenous, Black, and queer-of-color politics.

On my latest (though certainly not last) trip to the watershed with four of my students, two Watershed Council members put us all up in their houses and made us French toast at six a.m. to get us through a cold day of stream-crawling. The Watershed Council organized a spaghetti dinner for local residents, whom I knew from collecting water samples from their wells and springs. I presented results from a groundwater study analyzing these samples to trace flow paths of water from the ridgetop aquifer to sanctuary reaches in the streams. I passed out color copies of some of the figures from the scientific paper. They gave me gifts for my baby, a teething toy and a blue hat the shepherd had knit from her sheep's own wool. Such are the reciprocities of queer relational methodologies.

Because Salmon Creek is a close-knit community with minimal outside regulation, discussion often leads—as it did that night of the spaghetti dinner—toward collective governance of shared aquifers and common streams. In my practice there, which has elements of queer ecology method, my collaborators and I sometimes felt Muñoz's hopeless hopefulness as we gained strength from collective work in the field and the bar to weather the storms of despair and climate chaos that unravels lifeworlds. This collective work is "a politics oriented toward means and not ends" that builds collective affects, ones that Roderick Ferguson evoked so powerfully in his 2018 American Studies Association President's Lecture with, "We know the storm is coming, and we have everything we need."[29]

When I've shared writing on seasonal cycles of joy (in spring, at salmon persistence to spawn and emerge from gravel) and grief (in dry years, at finding pools devoid of fish or with dead fry floating in fetid water), mostly in academic talks, sometimes in print, many have responded with their own stories of sorrow, worry, and unsettledness at witnessing beloved species' death or distress. The natural and the social are inextricably tangled, always producing each other through data and conversations about what those data signify and how people, as individuals and members of local communities, should respond. If you, reader, are a scientist, I invite you to make such queer relations part of your scientific projects. I wonder what happens when you, unknown but beloved reader, share your own stories of such relation.

Underflow 5

Affects and Ecopoetics Practice

Queer and trans, as affects and subjectivities, are both underflows to normal science. Often intersecting with Indigenous, Black, Asian American/Pacific Islander, and Latinx identities, queer and trans bodies and modes of thought have long been present in ecology but marginalized and sometimes suppressed or invisible. Ecopoetics offers a way to evoke and channel these affects and practices. One ecopoetics technique—field writing—is a way of making a relation with one's immediate surroundings: call that the field, the world, the phenomenal, the present, the real.[1]

Field poetics urge writers to put aside goals and frameworks and to let the world in. Core to this practice is reading a counter-canon centering Native, Black, Latinx, queer, and feminist poets and critics and reading poetry through works of environmental science and history. In this reading through, we field poets do not try to integrate the findings of these diverse methods but rather to hold them in conversation, mindful of their authors' standpoints and political commitments. Reading these poets and scientists, we also look for clues to their practices of writing in the field, and we talk about these practices on trips or in guest lectures. In writing sessions, we bring students to a stretch of shoreline, then plunge them into exercises that explore their own poetics of relation to local waters. We always write too. Ripples, splashes, close encounters with jetsam, and interrupting birds culminate in a polyvocal recitation of our findings. Outside of class, July Hazard and I have used field writing to generate hypotheses for ecological studies, to write poems, to reflect on histories of conquest and ongoing Indigenous lifeways as resistance to colonial violence, and to make theory

Figure I5.1. Laying out a transect near the Duwamish River during a Queer Ecologies workshop. Rob Anderson and Sallie Lau choose the location while Anna Tsing records observations. Photo by Mark Stone.

about queer and trans embodiment in more-than-human dimensions. To illustrate the queer potential that Hazard and I find in ecopoetics methods for field composition, I'll briefly describe two of our favorite field activities: transects and diffractions.

Ecologists use transects to sample diversity of species or landscape features, laying out a tape measure across an area and enumerating the entities of interest that the tape measure falls across. In our version, poets can use a transect to sample memory and perception, mixing the science practice of sampling along a transect with a poetry practice that Brian Blanchfield developed to explore how spontaneous, associational memories could accompany deliberate remembering, inspired by Joe Brainard's *I Remember*.[2] In our version, rather than recording discrete entities like mussels or willow trees, poets record edges. But what is an edge? It is not an entity, but rather a query or probing of a category and its boundary. A transect of edges is a perversion of the transect, and using the noting of edges as an invitation to memory, in a lineage of queer poets, further queers the scientific practice.

Diffraction is a way philosophers can sense and explore how t hings come to matter in relation, grounded in the quantum physics

Figure 15.2. Participants in a Portage Bay ecopoetics class ask questions with thrown rocks and record the apparatus's response. Photo by author.

experimental method. As Karen Barad describes in *Meeting the Universe Halfway*, reflection is for representationalist methodologies and is set up to look for homologies and analogies. Diffraction, by contrast, is for nonrepresentationalist methodologies and attends to specific material entanglements: "Crucially, diffraction attends to the relational nature of difference; it does not figure difference as either a matter of essence or as inconsequential: 'a diffraction pattern does not map where differences appear, but rather maps where the effects of differences appear.'"[3] Diffraction is everywhere along shorelines, anywhere ripples produced by two or more disturbances interact.

As a practice for generating field writing, the diffraction exercise complements the active "doing something" of the transect (writing, measuring, touching, drawing, grasping) with "doing nothing" (looking, asking, listening, waiting, releasing). Tossing rocks in the water while holding a question in mind: what results? Diffraction, reflection, and refraction all happen simultaneously as ripples spread from where the rock enters the water. Focusing on one of these processes and asking the same question again and again with different rocks cause different answers to emerge in a poet's mind. The apparatus in this experiment includes the rock tosser; the rock itself in all of its provenance and history of formation, movement, and erosion; the wind and water; the engineering history that carved a channel through the portage path and lowered Lake Washington by three meters, so that pleasure boats could pass through; the other people there; the geese that land on the water, squawking madly; and so on. To extend the analogy to Barad's description of a physics experiment, the plunk-splash of the rock is the diffraction grating slit that the beam passes through, and the question-as-tagged-to-rock-body is what gets diffracted through the slit. What does this do to the rest of the question that doesn't make it through the slit? Such questions help poets attune to the shoreline and to their place in this apparatus and structure the poems that they compile from the activity. As these brief accounts show, this ecopoetics practice is both a queering of method and an interrogation of relation by people who are together in the field. This approach insists on the co-presence of poetry, art, and science in an ever unfolding and iterative call-and-response.

To introduce the transect of edges activity, July Hazard gave these instructions:

> On your paper, you'll mark a line down the middle, making two columns. In column 1, record the edges along your transect; in column 2, record a memory that arises when you notice that edge. So, for instance, at fifty-five centimeters, here's an edge of a pothole, so you're writing in column 1, "A cliff's edge breaks asphalt away from asphalt" . . . and suddenly you remember your sister swinging in a swing, pumping her legs, with magic marker flowers she drew on her feet, so now you're writing in column 2, "I remember the flowers my sister drew on her bare feet so she could swing better. I remember how dusty her feet were." Don't force the memories, and don't dwell on them—just let them arise as they will and try to jot down one vivid sense perception from each, then move on back to the transect, and find your next edge.

Then, as the students worked, we gave these instructions: "Don't excavate the whole memory. Just grab the image that flashed up. Look for edges—one person might see edges that are different from what another person sees. Your memories can be anything! They don't have to be edges! The columns are totally unrelated to each other! Just take the memories as they come!" To introduce the apparatus activity, I gave these instructions: "At the water's edge, pick up a rock, hold a question in your mind, and decide on what process you're going to focus on (reflection, refraction, or diffraction). Then toss the rock in the water and observe the disturbance it creates, and listen for an answer to your question. Record both question and answer in your notebook. Repeat several times, with different processes of focus." For both activities, our final instruction was to assemble the lines from both columns, in an order that feels right, into a poem. As a closing practice, we all stood in a circle and invited each person to read one line from their record of each activity; these became our collective poems.

5

With and for the Multitude

Cruising a Waterfront with José Esteban Muñoz

> Brown commons is meant to signify at least two things: one is the commons of brown people, places, feelings, sounds, animals, minerals, and other objects. What makes them brown is partially the way in which they suffer and strive together, but also the commonality of their ability to flourish under duress and pressure. They are brown in part because they have been devalued by the world outside their commons. Their brownness can be known by tracking the ways that global and local forces constantly attempt to degrade them and diminish their vitality. But they're also brown insofar as they smolder with life and persistence. They are brown because brown is a common color shared by a commons that is of and for the multitude.
>
> —JOSÉ ESTEBAN MUÑOZ,
> "Brown Commons: The Sense of Wildness"

Infrastructure as and in Emergency

Just before the pandemic shuttered businesses and cascaded through supply chains, silencing the rumble of shipping, drilling, trucks, railcars, and airplanes, July Hazard and I co-led a Queer Ecologies tour near the Duwamish River mouth in Seattle. Like many rivers in the US West, the Duwamish flows from alpine headwaters in national parks and forests, through farms, towns, and tree plantations. It then traverses factories and rail yards and spills into a constricted estuary whose tidelands lie under shipping piers and warehouses, now transforming into shiny condos. That day, twenty-odd humanities scholars, scientists, and anthropologists peered down at a scrap of wetland—a tidal marsh restoration—behind a bridge deflection boom, then climbed ten flights of stairs up to a bridge overlook. We surveyed the expanse of concrete, using the Waterlines map as our guide to its vanished, sinuous delta.[1]

Immediately below, a tugboat pushed a barge bound for Hawaii. In the distance, cranes unloaded container ships, silhouetted against a shiny skyline bristling with cranes. Along the riprapped shoreline, pleasure boats awaited after-work sailors; barges held the raw materials to fuel that building boom and the city's refuse. Like the rivers and tidal inlets around the world that now feature such encrustations of rock and cement, the Duwamish River has become industrial infrastructure centered on the Port of Seattle. Active and derelict manufacturing sites retain toxic strata and viscous residues that mingle with the underlying marsh and mudflat and seep from the riverbed. We led the Queer Ecologies participants through an incongruous block of Croatian fishermen's houses to a small port-owned park and fishing access. As in all such interstices, mosses, weeds, ants, and other small beings make their lives amid litter and rail ballast.

We pointed out traces of ecological relations amid the rumble of metal grinders and rush of semitrucks. Hazard and I theorized such persistence in the face of the slow violence of settler-colonial emergency as a feature of queer ecologies, which are also frequented by queer humans cruising for sex or stillness. With Anna Tsing, our guest of honor, we thought of such places as simultaneously global, articulated with atmospheric circulation and shipping routes, and hyperlocal, evolving in relation to tides, weather patterns, and practices of Indigenous and settler management. Part of what makes these spaces queer, we proposed, is how they remain vital, with an energy unto themselves. Queer ecologies methods bring that vitality into focus.

Before that field visit, Hazard and I often talked with colleagues and students about these paradoxes of infrastructure and vitality, emergency and ongoing survivance. We visited these shorelines to practice ecopoetics and develop queer methods. With some clams, barnacles, and industrial detritus, we made images that reveal the relations I'm sketching here. We had scrambled down an eroding bank to a river edge below the bridge and walked through small patches of marsh, fenced off from goose predation, to a rocky verge. The tide was out, revealing pilings that supported a home for retired sailors. One of us started picking up detritus and looking through it at the world. We took photographs of one another doing this, talking about how detritus became apparatus. This was an apparatus in Karen Barad's sense of a phenomenon, brought into being by an experimenter, which refuses the Cartesian split between subject and object. In these photographs, the image *is* and *shows/makes clear/ reveals* different human and more-than-human intra-actions that work through this apparatus.

The apparatus that made this photograph includes the ring of barnacle- and mussel-encrusted rusty metal; the stance and orientation of the person looking through it; the focal length and area of the camera; the tide; the river flow; the international trade agreements, flows of capital, and market dynamics that shaped the configuration of hard infrastructure at Harbor Island, across the water; and the politics that renamed the Duwamish River a waterway—a shipping canal rather than a living river. Also part of this apparatus: all of the people, rivers, and infrastructures that led to the production of this book and the energies and technologies to materialize it in your hands. The apparatus also includes July's and my collaboration, over nearly two decades, to overlay science and poetry on ongoing emergencies of settler colonialism and Manifest Destiny projects across the US West. Most recently and immediately, this apparatus emerged from a prompt in our Ecopoetics along Shorelines class: to make an apparatus for sensing buried flows.

During the 2020 pandemic spring, I expected to spend a lot of time along the lower Duwamish, providing technical advice to a Duwamish Tribe shoreline restoration and trail project and supervising students who are studying floating wetland barges as salmon habitat. Instead, I've visited just a handful of times, continuing my ecopoetics practice by photographing shoreline sites through holes in objects I find there. I visited other sites vicariously, via videos my students took during their socially distanced fieldwork. As the shelter-in-place period extended global quiet, the local sounds of the Duwamish floodplain changed from clangor of industry and buzz of airplanes to the jackhammer clatter of workers doing emergency rebuilding of the West Seattle Bridge, which has fatal structural damage caused by deferred maintenance and increased traffic to downtown tech hubs. The new sounds and silences evoke emergencies fast and slow. For a while, little traffic crossed the web of bridges over the river mouth, but now, in July 2020, "normal" traffic chokes the neighborhoods of Georgetown and South Park just upstream. COVID-19 caseloads in the predominantly poor and immigrant communities of the Duwamish Valley result from ongoing structural violence: crowded housing, lack of health care, and chronic disease caused by polluted air, soil, water, and workplaces.

To face and resist injustice and structural violence, queer ecology practices of being-together-in-place and forging bonds of kinship outside of blood and species lines are balm against isolation. The pandemic revealed these practices as crucial survival strategies in the face of state devaluation of Indigenous, Black, queer, trans, Brown, and immigrant lives. To think through queer kinship and ecologies in relation to local and global emergencies, I bring José Esteban Muñoz's posthumously published *Sense of Brown* down to the lower Duwamish.[2] The Duwamish is a river become infrastructure that embodies senses and qualities of the brown commons he invokes. To think in and as a brown commons, along wounded yet uncannily persistent waters, can inject radical politics into river restoration projects and disrupt white supremacist environmental discourses and practices that have long dominated urban shoreline management.

On Queer Method and Genealogy

Toxins and flood-channeling walls are also sometimes substrates for new life (barnacles, copepods) and human lifeways (fishing spots, cruising spots). In the late '90s and early 2000s, I and other genderqueers, dykes, and trans people brought dates to derelict in-between spaces along the shores of San Francisco and Oakland, shot porn, held fashion shows, and screened films there, and collected toxic detritus for collages or bonfires. Gay male cruising spots in parks and bathrooms were off-limits to us, or not where we would find our objects of desire. Along the waterfront, I could just be, a solitary queer or one of several freaks, anonymous. I belonged by dint of being there, how I moved through the space—looking or not looking, through gesture. Along these waterfronts I found ways of being in common—to not trample others, while not trying to do or make anything permanent or stake property claims.

In this chapter, I recount how I took Muñoz's idea of the brown commons down to the water, traced what happens as it tumbled downstream to where the tide washed it back and forth for a while. I sat on the beach. I felt how the brown commons might change if I laid it by the body of a spawned-out salmon, if I left it a while to grow barnacles. Like a boatyard raccoon that takes its nightly dumpster score down under the dock to wash in the brackish water, I dipped and rubbed the thoughts and feelings of Muñoz's text into the waters of the San Francisco estuary, the Monterey Bay, and the Duwamish River.

To explore brown commons—which are queer ecologies as well—as a source for environmental justice thought and action, this chapter takes as its method the hallowed queer practice of cruising postindustrial waterfronts.[3] This cruising is like and unlike Muñoz's approach in *Cruising Utopia*—a "kind of politicized cruising" that reveals the "anticipatory illumination of the utopian."[4] Muñoz reads queer performance archives across a felt sense of being together in queer spaces like performances and clubs. I read (not queer) river management and urban planning archives across Muñoz's queer texts and my memories of queer waterfront encounters. We both use practices of cruising—walking, noticing, desiring the sensations and matter of the city—as a way to bring these disparate entities into relation.

The waterfront—many specific waterfronts—well up into my text as memories, photographs, video stills, lists, and poems. Just as crucial as cruising methods to this chapter is the word *commons*, a word that is its own plural, a verb (*commoning*), and a term that retains its sense of the everyday, mundane—or as the old Homo Cult sticker said "Queer as fuck, common as muck." The awkward multitudinousness of commons—like sparkly broken glass on the edge of the water—displays differences among jetsams that are key to this chapter's proposition. In this praxis, brown commons is an apparatus that scientist-activist coalitions can deploy to transfigure a phenomenon, an urban river.

The brown commons is an idea and a text that I access as a trace. When José Esteban Muñoz passed in 2013, I had just taken my qualifying exams for my PhD and was following my partner, the poet July Hazard, to New York, to write for a semester. I had not yet encountered Muñoz as a scholar of the commons or queer utopian thought; friends introduced me to him on online memorial sites. We passed like someone you see across a club who never meets your eye. I came to know him, like so many queer writers who passed in the '80s and '90s, through writing and remembrance by his friends and queer family.[5] Such are queer lineages of inheritance and memory, poised outside the rush and grind of the everyday by time that acts as distance. I wanted more—to be in the room during a talk, or on the waterfront together, or drinking wine in his apartment salon.

I finally read Muñoz's *Cruising Utopia: The Then and There of Queer Futurity* in a different salon, one queer punk writer friends made—an old decrepit boatyard building on the Oakland waterfront. A handful of us gathered there every two weeks, barn doors open to the chill summer breeze, my green canoe hanging from the ceiling overhead. We read a fragment of the brown commons text that had just come out in the *GLQ* Queer Inhumanisms issue; the commons it evoked in my thought were urban commons, adjacent to water.

I wanted to read queer theory as a counterpoint, or perhaps counterweight, to the social and natural science that infused my PhD research that I had just completed. We traced a meandering canon through Muñoz's queer Latina foremother Gloria Anzaldúa and through other queer-of-color theorists whom I've engaged in this book.[6] Someone found a video of Muñoz and Samuel R. Delany reading on queer utopias: the opening lines of what became *The Sense of Brown*. We were following a trace, together in the undercommons.[7]

Fred Moten and Stefan Harney called the undercommons a dance floor.[8] If the brown commons is also a dance floor, I'm the wallflower leaning against the wall. I articulate a brown commons that is in relation to Muñoz's but that foregrounds riverine relations and fishy migrations that are outside his concern. Without conflating the brownness of racial formations (Muñoz's central interest) with the brownness of industrial afterlives (the new concern I lay alongside his), I explore both these valences of brown as animating what waterfronts do. *Do* here invokes what performance and queer theorists call performativity—with senses of actions performed and the feelings or actions that flow to those who witness them. Waterfronts' actions could be floods or barnacles growing or dredging and silting in channels. Those who witness could be human or not.

I began work on this project, in 2016, from then-extant outtakes of Muñoz's text: a videorecording of an event called Queer Utopias, featuring Muñoz and Samuel R. Delany in conversation and the short essay "The Sense of Brownness."⁹ I watched and rewatched the video of Muñoz reading "The Brown Commons: The Sense of Wildness," trying to catch and transcribe his words despite the crappy sound. I attuned to his movement: a cadenced back-and-forth while shifting his weight from foot to foot, switching the stack of papers from left to right hand and back to left, pushing his glasses up, stumbling over nested clauses, then smiling at the audience before starting in again. Searches for more fragments turned up Muñozian sources—posters for events, commentary on his talks, snapshots. Muñoz would often visit clubs or streets where the performers he studied lived or worked; I visited sites at the university and along the industrial waterfront where I would have liked to talk to him. When more texts and videos appeared on the internet or in journals, I cruised them too and found answers to some questions I had posed in drafts—on Althusser's swerve and chance, on how plants, animals, and waters figured for Muñoz as commons members, on how brown commons destabilize affects of whiteness, and how Muñoz's collaborators are thinking through the potentials and pitfalls of the commons as a concept in relation to gentrification, ecological crises, and urban space.[10]

These methods of cruising underflows of queer ecological thought give rise to this chapter's structure of interlinked sections: I first engage brown commons' resonance with environmental justice, their affinity for punk rock encounters and affects, their overlap with postindustrial urban waterfronts as social spaces, and their existence as queer ecologies. I then explore how the brown commons might engender new river resurgence strategies, using Seattle's Duwamish River as a case. I reject "community consultation" and instead propose a practice of grounded relationality for settlers, like me, who work in alliance with Native nations and frontline communities.[11] I close with a manifesto that channels "a shared affect of indignation that could lead to a thinking and analysis that would help assemble a self-conscious and potentially insurrectionist common."[12] This indignation, I imagine, churns up the placid stagnation of white and bureaucratic processes that govern Superfund cleanup and "neighborhood revitalization" along the Duwamish.[13]

Queer Ecology and Environmental Justice: The Endeavor of Knowing Brownness That Is Our Commonality

> Nonhuman brownness is only partially available to us, through the screen of human perception. But then, everything that I am describing as being brown is only partially knowable. To think about brownness is to accept that it arrives to us. We attune to it only partially. Pieces resist knowing and being known. At best, we can be attuned to what brownness does in the world—the performativity of brownness and the sense of the world such performances engender.[14]

Muñoz's commons are brown in at least two senses, as he explains. First, they are a more-than-human commons of "brown people, places, feelings, sounds, animals, minerals, and other objects"; second, they are tended by "brown people . . . who are rendered brown by their personal or familial participation in south-to-north migration patterns" and by their exclusion from white settler legal and property regimes by "the contempt and scorn of xenophobes, racists, and a class of people who are accustomed to savagely imposing their will on others." Reading "The Brown Commons: The Sense of Wildness" onstage, Muñoz channels an old-time anarchist soapbox agitator, in his insistent and driving cadence that lands again and again on the round vowels of "brown" and in the language of struggle. Brown commons "suffer and strive together"; they "flourish under duress and pressure"; they "smolder with life and persistence"—all in the first thirty seconds of the talk.

Waterfront spaces hold industries and their contaminant legacies. They often traverse Black, Latinx, or immigrant neighborhoods, Indigenous villages, and fishing spots. They are cruising or living places for queer and trans outcasts. Under the guise of ecological restoration and redevelopment, corporate and settler government actors design these waterfronts without grappling with their brownness or their status as commons. They describe industrial corridors and crumbling docklands as empty and valueless, just as their predecessors saw swamps and marshes as empty wastes.[15] They defend floodplains' and tidelands' transformation into shining office parks and luxury housing as creation of value. Ecological repair projects benefit some commoners, by improving fish harvests or shoreline access, while constraining others' life chances, by stepping up policing of cruising or sleeping outside.

Thinking waterfronts as brown commons/queer ecologies offers a way to understand such tensions as they play out along Seattle's Duwamish River. There, Muckleshoot commercial salmon fishermen span the river with nets to catch salmon. They make contact with other Native, Latinx, Black, Lao, Filipinx, Vietnamese, and white anglers fishing for subsistence or sport. All these fishers make distinct relations with the river and its fish and shellfish in spite of and also in resistance to the shared harm toxins do to their bodies and to the river's watershed body. Quoting Muñoz, such shared harm and refusal defines brownness: "Brown, it is important to mention, is not only the shared experience of harm between people and things; it is also the potential for the refusal and the resistance to that often-systemic harm. Brownness is a kind of uncanny persistence in the face of distressed conditions of possibility."[16]

With brown commons, Muñoz theorizes a shared harm without collapsing differences of racialization, Indigeneity, or class. Instead, this shared harm sometimes "activates being in-difference"—Roderick Ferguson's gesture to Avery Gordon's abolitionist concept of "a political consciousness and sensuous knowledge for living on better terms than we are offered, as if you had the necessity and freedom to do so."[17]

Ferguson, Muñoz, and their coconspirators advanced queer-of-color critique; Ferguson's plenary at the 2018 American Studies Association conference evoked a Black radical tradition that works in solidarity with Brown radical traditions, which Muñoz describes as one origin point for the brown commons. Sitting in an Atlanta conference room as a fabulous and jovial crowd assembled for Ferguson's lecture, I almost thought I saw Muñoz leaning along the wall and nodding in time to Ferguson's finale:

> In like spirit, [Vincent] Harding says, "I think part of the responsibility of Black scholars is to remind themselves and the community that they have constantly moved through darkness into light, constantly moved through pain into healing." We can go further and read this advice for all peoples and communities, characterized as they are by a vast reservoir of sources and powers. Engaging those sources and powers is our only chance to use the land in the way that the land intended: as the basis for envisioning and activating what Stuart Hall referred to as

"the interrelatedness of issues which is alone able to make the démarche from existing society to any other possibility."[18]

In thinking across Black feminist traditions and Indigenous and Brown land struggles as "sources and powers" for the renewal of the land, Ferguson evokes a radical strategy grounded in solidarity across difference.[19] By drawing together Muñoz's queer brown thought on the commons with environmental justice politics along urban rivers, I want to explicitly activate such sources and powers for renewal of lands and waters.

Though I don't think Muñoz planned to write in environmental studies, his thought is vital to environmental studies and activism. Brown commons that are queer ecologies are an underflow to environmental justice movements, showing them other ways to move, to shimmy, to take on capitalism, to be outrageous. This underflow brings feeling into discussions of species extinction and unavoidable sacrifice, cracking open flat bureaucratic affects with queer-of-color critique. Think, for example, of the term *brownfield*—policy-speak for contaminated urban land that must be remediated before it can be redeveloped into tax-generating property. Muñoz evokes a brown commons that refuses such logics of land and water as resources and also refuses the equation of brown with bad, contaminated, and in need of purification. He refuses to oppose human life with a nonhuman, nonsentient environment, instead centering connection and shared lifeworlds through cascading descriptions of how singularities make relations across species and kinds. This way of relation-making jibes with how a transfigured watershed body can bring beavers, salmon, humans, and other watershed inhabitants into renewed reciprocity. Take, for example, this passage: "People and things in the commons I'm rendering are brown because they share an organicism that is not solely organic of the natural as much as it is a certain brownness which is an embeddedness in a vast and pulsating social world. Again, not organic like self-sufficient organism, but organic in the way the world is brown and objects within the world touch and are co-present." The formulation "not solely organic of the natural" opens a space for lively vitality within "unnatural" landscapes like brownfields. Muñoz's theorization of the organic as a circuit among the social and the natural matters for a reparative ecological politics for at least two reasons.

First, *organic* at its root means carbon based. In river sciences, carbon is a tracer for flow through soil and water bodies. Carbon is a common substance composing people, fish, barnacles, trees turned into pier pilings, diesel fuel, PCBs, and other waterfront materials; some of these forms cause or hinder life processes. The food web itself is one carbon cycle; the subdiscipline of ecosystem ecology analyzes the linkages between local carbon and nutrient and global cycles—including how industrial emissions create feedbacks among systems. To think the organic with Muñoz is thus to think the ecosystem as a domain of the brown commons. Reworked through Muñoz's idea of "brownness which is an embeddedness in a vast and pulsating social world," ecological relations along the urban shore might unfold through a study of carbon's movement through, and transformation in, human and nonhuman bodies. For example, carbon—in the absence of oxygen—fuels anaerobic microbes that transform mercury into the harmful methylated form. Mercury accumulates in aquatic food webs as clams and fish eat plankton and are eaten in turn. If such scientific analyses began from "the continuous co-constitution of life-worlds between humans and others," as Métis scholar Zoe Todd writes regarding oil relationalities, they might generate novel perspectives on urban contaminants and their transformation.[20]

Second, Muñoz's queer brown thought matters for ecological science concerned with material interactions (or intra-actions) because organicism transgresses natural and social realms and requires reciprocity.[21] "Not organic like self-sufficient" centers relations among people yet extends a certain agency to "objects" and "the natural" to influence culture and politics.[22] Muñoz invokes a lively feminist materiality. He takes up Jane Bennett's "grand concordance of things" but rejects assemblage theory as too placid: "Life in the commons should be turbulent," always resisting "various enclosures that attempt to overwhelm a commons." Human members of the commons stir up turbulence as they work to activate and stimulate commons' "insurrectionist promise." Ecologists recognize the lively materiality of molecules, waters, plants, animals, and forces of nature, and they excel at modeling and imagining turbulent and dynamic cascades of events that emerge through their interplay.[23] Yet many ecologists would stop short of imagining such turbulence and dynamism within the realm of politics as of a kind with ecological cycles. That is, they would dismiss Muñoz's argument as metaphorical. That ecology holds insurrectionist promise finds little purchase among ecological scientists, at least most I've met. As a queer ecologist who travels in such promise in street protests and organizing spaces, I hold both valences of turbulence and liveliness as real—and also powerful as metaphor: a transfer of thought from one thinker to another. I use queer ecological thought to render insurrectionary promise possible.

The Central Swerve of the Encounter Is a Collision That Can Be Turbulent or Harmonious

Roiling in such turbulence, what kind of concord might one then find in the brown commons? Muñoz does not say, at least in texts that were extant at this writing. But in the 2013 "Gimme Gimme," Muñoz follows the poet Jack Spicer down to the waterfront and comes to concord from a different slant.

> "Indefiniteness is an element of the true music."
> The grand concord of what
> Does not stoop to definition. The seagull
> Alone on the pier cawing its head off
> Over no fish, no other seagull.
> No ocean. As absolutely devoid of meaning
>
> As a French horn.
> It is not even an orchestra. Concord
> Alone on a pier. The grand concord of what
>
> Does not stoop to definition. No fish
> No other seagull, no ocean—the true
>
> Music.[24]

Muñoz riffs on that poem, and in particular "what / Does not stoop to definition" to describe commons created by and among punks (including many young people of color) in the early LA scene. Spicer's image of the seagull "alone on the pier cawing its head off / Over no fish, no other seagull" is a performance of queer excess for no audience or purpose that nonetheless evokes and animates fish, the ocean, the queer pier. Muñoz maps this more-than-human excess of Spicer's waterfront scene onto an explication of the brown commons' performativity—what is done within the brown commons to bring it into being, and what it does by becoming: "Punk social relations can be thought of as aleatory and belonging to the realm of the clinamen. The central swerve of the encounter is a collision that can be turbulent or harmonious, an essentially incalculable concord of that which does, in Spicer's phrasing, not stoop to definition. The crash of atoms becomes the crash of bodies in dark spaces, the willful unharmonious crash of chords and notes, and in the early days of one particular punk rock scene, the clash of young people from different social geographies converging in dank little clubs in Hollywood."[25]

Punk rock commons are one origination point for the brown commons; another is the Latinx drag bar the Silver Platter in Los Angeles, the central character in Wu Tsang's film *Wildness*, which grounds Muñoz's exploration of brown commons.[26] Both are spaces where Brown punks and queers perform and create excess—excessive gesture, excessive noise, excessive fabulousness—in a negation of the enclosure and limitations that dominant society imposes outside the walls of those commons. Such excess could be (though Muñoz doesn't say as much) a stand-in for infinite other gestures and performances of excess by trees, wantonly shedding their leaves, or salmon, excessively leaving their bodies for bears and raccoons and feral dogs to drag off. Excess is the basis of ecology and the nonhuman world that capital encloses. By recognizing more-than-human excess and the futility of hoarding anything, humans can join with the nonhuman to make brown commons. To do so is to negate logics of enclosure, of Manifest Destiny.

Muñoz writes of such rejection, "The brown punks and queers discussed here . . . perform an insistence on wanting more in the face of scarcity, which does more than simply reject negation, but instead, works through it to imagine a being-in-common within the negative."[27] What does it mean to oppose negation with concord, and how does either concept relate to politics of environmental justice and land renewal? The brown commons' punk scrappiness and emphasis on touch and co-presence feels right at the edge of the water. There, tugboat waves lap on stones that, on closer examination, are rounded chunks of concrete and jagged scraps of slag embedded in mud. Affect—which Muñoz characterizes as an intensification of what it is to be brown—amplifies through belonging to or inhabiting, through being embedded, as a clam in tideflat mud, in a "vast and pulsating social world." The feeling is broad, yet intimate, like a city is, or like the feeling of dancing among strangers at a really good show.

To think about waterfront social relations through the brown commons is to reject narratives of inevitable toxicity and damage—an act of negation. Through this negation, brown commons make an ecological concord—turbulent though it may be—among singularities (human and otherwise) that become multiple through their desire for one another and their ongoing relations with water. In the face of local and global harm and damage, this desire is utopian, a not-yet-here queer futurity rooted in history. Muñoz's attention to politics of race, migration, and capital's traverse of national boundaries also fits along the Duwamish and other port waterfronts, which greet immigrants and travelers and move goods onward in global market circuits.

Such desire and past-present-future imagination resonate at the intersection of queer-of-color critique and decolonizing methods. Writing of desire as a refusal of colonial narratives, Eve Tuck and K. Wayne Yang write, "Rooted in possibilities gone but not foreclosed, 'the *not yet*, and at times, the *not anymore*,' desire refuses the master narrative that colonization was inevitable and has a monopoly on the future. By refusing the teleos of colonial future, desire expands possible futures. As a mode of refusal, desire is a 'no' and a 'yes.'"[28] This "no" and "yes" of refusal rejects binary thinking that undergirds white supremacist narratives of progress; this progress degrades and contaminates, yet life and vitality persist in the face of this contamination. Examples of such refusal and desire abound in the brown commons along the Duwamish. Among people who spend time along the river, the desire to eat fish they catch from the river and to bring their children there to play refuses labels of intrinsic or already finished damage. This desire foregrounds community capacities and desires that support the collective continuance of immigrant and Indigenous communities.[29]

In the 1970s, Muckleshoot tribal fishermen's refusal of racist Washington state fishing enforcement sparked one skirmish in the Fish Wars. Those protests culminated in the Boldt Decision, which affirmed treaty tribes' right to comanage fisheries and to reserve 50 percent of the harvestable catch. Local residents and Muckleshoot and Duwamish tribal members, working through the nonprofit Duwamish River Cleanup Coalition, fought to design parks, greenways, and fishing access into Superfund cleanup.[30] The coalition now also runs green jobs training programs and youth environmental science projects, and it campaigned to rename Seattle's "T-107 Park" Ha-Ah-Poos (*he?apus*)—Duwamish Village Park, to keep present the village once located there. These examples contain both the no and the yes of refusal.

Tuck and Yang do not discuss nonhuman animals' potential refusals of research. Their argument rests on voice, narrative, and story and so cannot be extended to animals, who don't make narrative. Nonetheless, Tuck and Yang's third axiom, "Research may not be the intervention that is needed," arguably applies to environmental cleanup projects that aim to increase life chances for these species, for which abundant scientific evidence and documentation already exists. They write, "In cases in which an intervention is needed, there are many other ways of developing and communicating ideas, including billboards, blogs, bumper stickers, letters, compelling spokespersons, flash mobs, YouTube videos, curricula, open houses, community talking tours, postcards, and the many forms of art."[31]

In new work cogenerated with the Duwamish Tribe, I am convening a group of University of Washington professors and students who work with Duwamish Valley communities into a research coordination network. This network responds to the Tribe's and environmental justice organizations' priorities and channels research funding and resources to local residents. Responding to Tuck and Yang's provocation, the pilot project includes community history talking tours, a community science mapping and species identification project, and events focused on generating and sharing creative writing and art. We test storm water for waste seepage, watch for birds and salmon with houseless folks living in parks and campers, and fly drones with youth corps kids to see the watershed in a new way. This project, convened by BIPOC and queer scientists and geographers, is its own brown commons. Our spatial and scientific work focuses on a small tributary that flows near the Duwamish Longhouse. It should empty out into the Duwamish in a cruising park but instead enters a culvert and is pumped and rerouted downstream, through a different outlet pipe. As we trace these underflows, we are feeling around for stories, histories, and desires with community scientists.

The brown commons' challenge, in such research projects, is to transfigure research into a more just and relational form. Midway through my work on this chapter, I was invited by landscape architects and ecologists at the University of Washington's Green Futures Lab to advise on science and community engagement for a floating wetland design project. The science of making riprapped shorelines into "living shorelines" could be seen as a way of measuring a singular fish's desire to become multiple, by aggregating on the floating wetlands. I came into the project late, after the first grant was funded and the barges were planted and almost ready to launch. I contributed peripherally as an adviser to master's students collecting field data with local residents working as community scientists. Following straight science's norms, the students observed how fish and birds reacted to the structures, made water quality measurements with community scientists, and worked with school kids from across the city to harvest plants and bugs that grew in the wetland's substrate of layered coconut-husk matting and straw. Paying community members to work a few hours each week, to help count fish and measure water, was an important first step. But to do such projects in and as a brown commons requires more of researchers: that we work with community members from the outset of such projects, to center their desires in the scientific investigations and architectural designs. This is the commitment of research cogeneration and environmental justice: to center the knowledges and practices of those disproportionately harmed by extraction and pollution, by amplifying knowledge they produced through "uncanny persistence in the face of shared wounding."

One of the student researchers in my lab, Charlotte Dohrn, reflected on this experience in my and July Hazard's ecopoetics class, by creating a sculpture/performance called *Substrate*, a miniature floating wetland that she released into the waters of Portage Bay. I read her writing about the justice dimensions of artwork as a clash and swerve of an encounter with the river and a foray toward relation with the politics of the brown commons. She writes,

> Relocating a few bulrushes is the physical result of this project—
> perhaps they will float away and take root somewhere nearby. . . . The

process of gathering [them] helped me reflect on the physical places and traces of the shorelines where I spend time.... The floating wetlands were part of an imagined future vision of abundant juvenile salmon traveling downriver, sheltering and feeding under lush floating wetlands affixed to a few remaining bulkheads between soft shoreline.... *Substrate* [indexes the physical] substrate beneath wetland plant and animal communities, the vast amount of work ahead to clean and contain the contaminated substrate of the Duwamish River, the layering of shoreline relational art on top of the floating wetlands research project, and the feeling of returning to home to a city that is both familiar and unfamiliar.

As I imagine it, the brown commons as a framework for waterfront research encompasses action research principles, then exceeds their boundaries by making queer insurrectionary moves against capital's enclosures.

[The Waterfront] Engenders a Certain Kind of Encounter, a Weird Clinamen, Where Matter, Sound, and People Collide

Urban waterfronts are mosaics: of industrial activity, postindustrial neglect and contamination, sanitized redevelopment, and public or privatized green space. Overlain on them are the swoop, dive, and hustle of people, birds, rats, crabs, microbes, plankton, and fishes. Like the swerve and tumble of sweaty bodies in a punk club, the shoreline makes a border that is porous, tidal, contingent, and constantly becoming. Because of this, of all the urban spaces, waterfronts feel the more-than-human most. Or if a watershed cannot be said to feel, exactly, at least one can say that feelings collect there. They circulate with human bodies-in-contact. Perhaps, following Muñoz, watersheds can be said to engender feeling: "We know that some humans are brown, and that they feel differently. Things are brown in that they radiate a different kind of affect. *Affect* as I am employing it in this project is meant to address a sense of being in common as it is transmitted across people, place, space. Brown affect traverses the rhythmic spacings between those singularities that compose the plurality of the commons."

How exactly does brown affect flow between more-than-human singularities? Might a beach or pier absorb the feelings of the people who come there to fish or stare at the water? If fish feel, do their feelings lap against the rocks with the waves? Or, like punk clubs and the Silver Platter, do shorelines rather engender encounters, which in turn provoke feelings of different kinds? Waterfronts do generate affects of negation that can lead to being otherwise in common, even on spruced up tourist drags like Seattle's Elliott Bay or San Francisco's Embarcadero, and especially where marshes accrete amid jagged pilings and crumbling levees. These waterfronts are homes and sources of livelihood for Indigenous, immigrant, and unhoused people; they are cruised by queers and sex workers and fished by everyone. They are constantly traversed by flows of water, sediment, and aquatic life. To cruise these spaces is to attune to the ways that these waterfronts are brown and queer, and to think them in opposition to colonizing or gentrifying impulses that would fill in tidelands and floodplains, tame floods and channel tides. The colonizing impulse tones down exuberant life in the ever-shifting littoral.

This exuberance, this queer excess, characterizes living things but also inanimate, sometimes toxic releases. On the Duwamish, these include creosote from piers, PCBs from shuttered power plants, sewage from overflow pipes, metals and acidity from abandoned mines. Such inhuman sources' animacy, Mel Chen has shown, destabilizes hierarchies of race and sexuality.[32] Cruising the Duwamish in search of its specific histories and futurities of resistance to enclosure, I feel other moments of waterfront resistance that are underflows to the relentless redevelopment and gentrification that has eradicated such places. I'm thinking here specifically of South San Francisco, the Oakland waterfront, Dead Horse Bay and the Gowanus Canal in Brooklyn—places that I have cruised as often as I can. Tides, slimes, and encrustations rework slag and cast-off objects into strange forms. Moments of protest against power plants and ineffective toxic cleanup linger as feelings that resonate with the ongoing transformation wrought by plants and shellfish, before restoration plans impose fixed elevations and land-water boundaries.[33] This ongoing transformation is exuberance and excess that I figure as more-than-human resistance.

Such exuberance, uncovered while cruising in the trace, characterizes environmental justice protests. Environmental justice movements share lineages with the Brown, Black, Red, and Gay Power '70s movements that Muñoz invokes as lineages in the brown commons. Environmental justice politics, as one through-line in these movements, "attempt to articulate a refusal of dominant logics and systems of thought, and [to create] systems of thinking and doing otherwise." For Muñoz, turning to such radical histories "makes a point that the world is not becoming brown, but has been brown." Turning to history is a crucial practice for young trans activists like *Wildness* director Wu Tsang and a younger me, who are trying to project a desire for brownness and commonness into a more livable future. Environmental justice marches and encampments, beginning with Love Canal and extending through recent protests at Standing Rock, in coal country, and for climate justice, have enacted refusals of dominant logics of extraction and pollution. Participants in these movements build "systems of thinking and doing otherwise" through everyday actions of fishing, holding ceremony, gardening, and performance. This motion and movement, from insurrectionist blockade to everyday commoning work, is all part of the work Rod Ferguson describes happening "in the gap between what we need and what we have."[34]

Let us thus think through the waterfront's past in order to specify these dominant logics. Dredging and rerouting streams and tidelands were—and continue to be—Manifest Destiny projects, and the idea that these projects can improve unruly commons are Manifest Destiny logics. The brown commons disrupts such dominant logics by contesting the idea that shoreline engineering projects are inevitable and permanent. The brown commons instead sees urban water edges as a past-present-future call-and-response that constantly reshapes its boundary. Riprap may seem for a while to fix the shoreline, thereby facilitating real estate speculation and extraction of value. But eventually, the river or the tide will undermine some armoring or overtop a levee. All along, people—principally from Native nations, but also other fishers, immigrants, and workers—have rejected capital's sacrifice of more-than-human relations for settler "progress." River creatures themselves have persisted, although much reduced in number, long enough for new alliances to respond with toxic cleanup and habitat creation.

Following Muñoz, thinking the wild alongside the brown can reveal the resulting breach in the fixed bank.[35] This breach makes a fractal, crenellated edge called a living shoreline along raised and filled tidelands, as an emergence of a brown common. To take one specific example, along Seattle's Duwamish River, where restoration projects are increasing life chances for salmon, birds, and other creatures, this urban waterfront is not becoming wild, but has been wild. It is not becoming ecological, vibrant, inhabited, or used by people but has always been so—in a call-and-response that only partly ever follows the dictates of engineers that build these infrastructure projects or the planners that seek to remake them through ecological restoration. What if these engineering and planning processes became processes of attunement to the brown commons that has always been?

The Queer Ecology—the Organic and the Inorganic— Which Is the Brown Commons

In what sense are brown commons queer ecologies? Muñoz's core concern is not ecology as in ecological science.[36] Rather, I see Muñoz as animating the ecological as metaphor to evoke the vitality, and I would say stochasticity, in urban clubs and street scenes. In his discussion of Wu Tsang's documentary *Wildness,* Muñoz shows a clip of Los Angeles streets at night, shot out the window of a car that arrives at the Silver Platter bar, then he describes the liveliness of the brown commons the film brings into being:

> A mysterious voice has been narrating to heavy and ominous drum score. In Spanish, the voice describes how time is borrowed, how it changes everything, how there are not many like her, and what will become of her. Viewers soon realize that the throaty feminine voice has been assigned, not to a person, but to a place, a place that holds and shelters brown life within its walls. . . . Long before humans take part in the film's narrative, we witness a commons of lights, shadows, commerce, buildings, trees. It's not until a brown commons is visualized as street, ambience, light, movement, that the documentary's protagonist can come into view. Viewers encounter the bar's glimmering sign, its luminous, old-school beauty suggesting another time and place throbbing cumbia, as the space erupts with dancers.

Although Muñoz does not define the queer ecologies that are brown commons as solely urban phenomena, his sites are urban: performance venues and the streets, parking lots, and late-night cafés where punks and queer artists congregate after hours. What emerges from the unpredictable encounters in and around the Silver Platter is a feeling and politics that exceeds the human, and, for me, evokes the ecological concept of emergence.

Both Tsang, in shooting and editing the opening sequence, and Muñoz, in reading it, employ a cruising method that is of the body and the sensorium, whether attuning to performance in person or archival footage, in moving through the city itself, or in imagining the queer genealogies that connect their movements through the city across time.[37] This cruising, not for sex but for queer feeling and vitality, structures Muñoz's writing about the New York performance scene, as when he walked alone from his apartment to where the dancer Fred Herko lived in the 1960s or, with collaborators Jack Halberstam and Tavia Nyong'o, devised a special issue on wildness on "a series of summer walks into the urban wild."[38] From such queer encounters, Muñoz theorizes a sense of liveliness, of contact, vitality, the multitude, co-presence, "a movement, a flow, an impulse to move beyond singular individualized subjectivities."

Cruising as method is rooted in a desire that includes and exceeds sex and intensifies everyday encounters such that objects "smolder with life and persistence."[39] I cruise waterfronts and streams in a way that feels similar. This cruising practice is an underflow that relates to and can transfigure urban ecological science and planning along shorelines whose inhabitants "suffer and strive together."[40] Such cruising is a queer underflow to field science practice in three senses: First, as a way of seeing and relating to tides, slimes, and encrustations of shellfish on industrial debris—queer ecologists and artists materialize queer futures in artworks and encounters. Second, as a theoretical orientation to queer affiliations made beyond heterosexual encounter and pair bonding—queer orientations to ecology and evolutionary biology research may give rise to new hypotheses for study and action. Third, as collective politics influenced by queer-of-color people and thought, against policies designed to exterminate queers and other outcasts—Black Lives Matter and Water Protector movements connect climate and water justice to demands for decolonization and reparations and enact queer futurity through mutual aid, urban gardening, and ceremony as a core element of protest.

Cruising waterfront queer ecologies with Muñoz brings into focus circuits of feeling that pass between people, places, organisms, and elementals. As an ecologist and radical queer, I see such languages and actions as negation of white affects and communicative norms that dominate Superfund cleanup, urban planning, river restoration, and other scientific-bureaucratic spaces.[41] Queer-trans Brown (and Black and Indigenous) affects circulate in these processes as underflows, at community river festivals, on fishing piers, and whenever people cruise the shorelines in search of unexpected encounters, human and otherwise. For Muñoz, queer ecologies are both "real space, actual space, and symbolic spaces and affective landscapes, the space of shared emotion and shared modes of feeling that cross space."[42] In the opening shot of *Wildness*, the camera takes the perspective of someone cruising in a car down Wilshire in LA, at night, past MacArthur Park's dark lake lit golden with sodium lights, through a dreamlike landscape of red and gold and white street lights, storefronts of taco shops, bodegas, a vibrant everynight streetscape.

Describing the opening shot of *Wildness*, Muñoz commented, "There are so many slow shots that aren't just . . . about establishing the location of the bar, but really about the city itself. [Evoking] the bar's place in what I'm calling this queer ecology, which is an urban ecology that includes the natural and the unnatural, the organic and the inorganic." Muñoz's commentary on this sequence extends Jean-Luc Nancy's concept of being with, of an I that is not prior to a we, beyond humans to animals, streets, buildings, and the night itself. These shots suggest that the city, the night, the park, are part of what goes on inside the bar, are of a piece with it through circuits of emotion that radiate from events that happen in the bar and outside in the world, including deportations, gentrification, and antitrans violence.

Against environmental studies' dominant white affects that excise emotion from scientific discourse, brown commons' insurrectionist tactics assert urban spaces as queer ecologies. Imagining, feeling, and being otherwise with the waterfront could shape ecologists' work to repair damage from the "violence of property, finance, and capital's overarching mechanisms of domination."[43]

Unsettling Moves in Brown Commons

Waterfronts are queer spaces in a specific sense. They are sites of formative queer sex and sociality. In the late '90s and early 2000s, by night I frequented, alone or in groups of punks and queers, a former dump become wetlands and thickets of fennel, and abandoned power plants and train stations, where we staged performances. We encountered other outcasts in passing, or as traces—commons that were vibrant tapestries of emergent relations—what Muñoz calls "the swerve of matter, organic and otherwise. The moment of contact, the encounter and all it can generate. . . . Being with, being alongside."[44] Now, years later along the Duwamish, alone or with other queers, I reenter this mode of feeling-in-the-world.[45] The brown commons I am describing are not "a stand-in for nature and the ways in which it is subsumed by private property"—they come to be in a city that is, as Muñoz writes, "not just a built environment, an object of capitalist consumption, but . . . a living dynamic of cultural practices, intellectual circuits, affective networks, and social institutions . . . not just as an enclosure of nature, but as its own common, that is teeming with the potentiality of a kind of living otherwise that a full engagement with the commons can help actualize."

In Seattle, settler land grabs displaced Duwamish villages along the eponymous river's tidelands and floodplains; an engineering project dried up its Black River tributary, precluding salmon harvest at the village of *sq̓ʷuʔq̓ʷuʔ* (Confluence of Waters). From Seattle to San Francisco to the San Joaquin Valley, and along thousands of rivers in between, settlers viewed wetlands as wastelands that could be improved into property only by corralling the dynamic rush and flow of water, sediment, and organisms behind levee walls. These property relations are constitutive of ecological relations within settler-colonial states. Property rights determine how people manage rivers and tidelands and constrain ecological study, cleanup, and recovery. I read Muñoz's "potentiality of a kind of living otherwise" as incitement to unsettling: a call to disrupt property relations that began with and perpetuate colonialism.

This reading of brown commons as a project of unsettling is a reparative one, a "critical practice that seeks to love and nurture its objects of study."[46] I speculate about how Muñoz might have engaged recent critiques of the commons by Indigenous scholars.[47] Déné scholar Glen

Coulthard's critique of the commons overlaps with Muñoz's assertion that brown commons is not just land or nature but includes "actual space, symbolic spaces, and affective landscapes, and the space of shared emotion and shared modes of feeling that cross space."[48] Coulthard said in an interview, "If we are committed to reclaiming the commons we are going to have to work critically to re-establish non-capitalist and decolonial social relations and legal traditions that have survived through generations of Indigenous communities. It's not just about land; it's about the legal and customary relationships that emerge from our connection to the land that are integral to imagining new formations beyond private property."[49]

Describing disruptions of the social relations of private property, Kim TallBear has written that genocide is "the simultaneous destruction of co-constituted peoples and their nonhuman relations."[50] This definition is useful in thinking across Indigenous and queer-of-color responses to settler attempts to undermine Indigenous, Black, and Brown relations along rivers. These rivers flow through urban and rural lands, where cities and companies dredge, fill, and poison shoreline and aquatic lifeworlds. Such projects interfere with treaty rights to fish, hunt, and harvest. Amid waste streams that urban waters accumulate, for a Muckleshoot fisherman to insist on treaty rights to harvest salmon is "a kind of living otherwise," which fosters collective continuance by maintaining social relations with people and other-than-human beings.

For the Duwamish Tribe, to build a longhouse near the old village of he?apus, set terms for Superfund cleanup, and restore shoreline habitats is an assertion of sovereignty, against US rescinding of formal recognition. For an immigrant or rural settler descendant, to maintain fishing traditions through urban subsistence harvest is another kind of living otherwise. So too are everyday and organized actions of stewardship and river cleanup. Such actions go unnoticed by dominant society, unless they spark oppositional movements. For example, Duwamish Tribe chairman James Rasmussen and white Vietnam veteran John Beale each worked for decades along the river to restore pockets of wetlands along the industrial shoreline. In the '90s, they began patrolling the river together to catch polluters, spurred the EPA to add the river to the Superfund, and eventually convened the Duwamish River Cleanup Coalition. Through such coalitional work, the Duwamish Tribe and the organizations that

John Beale founded work together with others who may not share their politics, and they build solidarity across difference by maintaining distinct engagements with the river and regulatory processes.[51] This coalition of different actors disrupts settler governments' organization and deployment of what Muñoz calls the "harsh asymmetries that systemically devalue classes of singularities,"[52] by framing river cleanup and restoration as direct action that engenders individual and collective healing.

The brown commons enables us to understand such acts as underflows. Muñoz describes how everyday practices and the swerve of chance encounters create a queer sociality that contests a normative relation to rivers under capitalism. This sense of queer sociality, as Muñoz makes clear, is not limited to sexuality or gender difference. Brown commons index many ways of living beyond the normal or sanctioned activities of urban waterfronts, while centering race and migration. Such underflows, whether they make Brown commons or Black or Indigenous ones, contest and negate extractive and destructive settler logics. They are becomings that maintain difference along lines of race and gender as people move "from isolated individual particularities to a better sense of the multiple."[53] I read Muñoz's move here as displaying decolonizing ethics and working in sync with Coulthard's project of "imagining new formations beyond private property."

Toxicity as Shared Wounding

> I mean a brownness that is conferred by the way one's spatial coordinates are contested, the ways in which one's right to residency is challenged by those who make false claims of nativity. I also think of brownness in relation to customs and everyday styles of living that connote a sense of illegitimacy. Brown indexes a certain vulnerability to mechanisms of domination.[54]

I learned an unofficial, word-of-mouth, oral history of toxicity from antipower plant activists and environmental educators in the Bay Area, and so I imagined a similar history's imprint along the Duwamish even before I learned its particularities. Dredges cut or buried marshy edges along waterfronts, creating land for flood-prone neighborhoods of dockworkers, immigrants, and people of color. Wartime industry's toxins magnified in the bodies of fish and shellfish that subsistence fishermen catch and eat, and highways and heavy industry continue to pollute air, land, and water. Though Oakland's and San Francisco's waterfronts started as a salt marsh and South Seattle's as an unruly glacial river, they look remarkably similar now: grim riprap walls bolstering massive cranes or cement plants, interspersed with short reaches of marinas, low-rise office complexes, and public parks.

Superfund designation mandates that toxic waste be removed or covered over and provides monetary compensation to remediate past damage. It does not address the ongoing toxicity that industrial infrastructure enables, such as storm-water overflow or air pollution from factories and highways.[55] Along the Duwamish and San Francisco Bay, environmental justice campaigns have become multigenerational processes that produce reams of maps and documents. Toxic temporalities index impacts to human and nonhuman bodies through time and long past the toxic "event" of power plant construction or a toxic spill: a queer temporality of waiting and persisting.

Residents' desire for their neighborhoods to be free from toxicity brings human and nonhuman bodies into relation along the Duwamish and other urban waterfronts. The traces of toxicity in these human bodies are uneven, and the river is not the only source of their body burden: though wounding is shared, it is also differential. People with less mobility breathe in more of the air pollution; those who rely on subsistence fishing and shellfishing consume more biomagnified mercury and PCBs.

At an urban wetland next to a power plant in San Francisco's Hunters Point neighborhood in the 2000s, sixth graders on a "toxic tour" I led related how human and fish bodies experience different temporalities of exposures. In their field journals and letters to the mayor, they wrote things like "If it is bad for us to swim in this water, what happens to the stripers that swim in it all the time?" "If we should only eat a few fish or mussels from the Bay, what about the fish or seals that only eat those species?" After a long campaign, local residents and allies won the power plant's closure in the mid-2000s, although PCBs and mercury persist in sediments. On the Duwamish in 2020, salmon, a transient species, is presented by public health officials as safe to eat (although "blackmouth salmon," Chinook that spend two years in the river, are best avoided). The temporalities of when salmon are available and when people fish on the river do not always overlap. For example, during seasons when salmon are not running, some subsistence fishers still fish and therefore consume species with a heavy toxic burden.

Public health campaigns are translated into cultural practice through annual water festivals along the Duwamish River. In 2018, at the largest gathering place in the South Park neighborhood, kids cheered for *lucha libre* performers and made biodegradable floating candle lanterns from rice cakes and banana leaves.[56] The Duwamish Waterway Park is a flat expanse of grass, a baseball diamond, and a small beach between tall boulder retaining walls. Between youth performances, chefs demonstrated how to safely cook Duwamish River salmon, Khmer and Mexican-Cajun style. A Filipinx drag queen narrated as a Lao man expertly fileted a king salmon, then sliced the PCB-rich fat off the belly and around the fins. She joked with the Latinx and Lao chefs about their secret spices as they cooked up the fish. Eating a fish that swam up our local industrial river while sitting on its banks, with the music and wrestlers yelling and kids splashing by the river, we were all drawn into a swerve of encounter with a river where salmon were abundant, safe, and delicious. Many visitors came back for seconds and thirds. My one-year-old kid tasted salmon for the first time and loved it and the taiko drum performance in equal measure. In making this future in the present, these activists invited the gathered river lovers to activate being in-difference, by continuing to engage in the Superfund cleanup, and by organizing against the gentrification that is displacing many longtime residents. This activation celebrated fishers' daily acts of becoming-with the Duwamish and brought us all into closer relation with the salmon and the river in lively resistance.

To Think the Inhuman Is the Necessary Queer Labor of the Incommensurate

With key potentials of waterfront brown commons in mind, let us now walk back down to the edge of the water, a postindustrial shoreline along the Duwamish, which may share characteristics with a waterfront that you frequent. This urban industrial waterfront is water you don't get into. The brown commons can show how to think with such waterfront spaces about becoming—the ongoingness of barnacles on the pier, fish in the water, Muckleshoot fishers catching them in nets and Latinx and Lao and white fishers catching them with poles, kids throwing stones into the water: "Brown commons are not made. One does not bring them into being. Instead we become attuned to them." On the Duwamish, as along other industrial waterfronts, people make affective relations with one another and with the river by responding to and resisting regulation, gentrification, and official planning. I theorize these relations as underflows to governmental processes of shoreline erasure and repair; Muñoz might call them traces.

Habitat restoration projects can be engines of displacement and gentrification or serve as greenwashing for polluters.[57] Duwamish Valley residents are acutely aware of this potential. Thinking as the brown commons provokes restoration and redevelopment practitioners to confront "uncanny persistence in the face of shared wounding." It is this uncanny persistence that, on many rivers and estuaries, has impelled Superfund listing, higher levels of cleanup, and repair of ecological function. The remnant oxbow and island at Herring's House was preserved from dredging by Duwamish Tribe members' blockade of bulldozers in the 1950s, and the Tribe's work to unpoison the river is part of ongoing resurgence.[58] Since John Beale began restoring Hamm Creek in the 1970s, thousands of volunteers have come to know and fight for the river through shoreline work with him. The Duwamish River Cleanup Coalition and other environmental justice groups have won funding for job training and a youth environmental corps and are fighting for affordable housing and clean industrial jobs as a bulwark against further gentrification and displacement. So far, all these efforts have created small pockets of green wetland amid concrete or riprap-lined banks. These scraps have become touchstones for river restoration workers who imagine expanding them enough to let the ecosystem make its own way. In response to these human actions, logs drift in and catch on engineered logjams, birds and crabs find these pockets of food and slow water carved out of parking lots, and microbes slowly transform toxins in sediments.

But while youths, workers, and low-income residents are included in environmental justice consultation processes, other people who help make these brown commons are overlooked or dismissed as worthless and trashy. Abandoned piers where riffraff have late-night bonfires, the shady willow grove where houseless people sleep and relax, sailboat squatters' illicit moorings and dinghy landings—the brown commons reveals such places as world-making projects by people "cast out of straight time's rhythms," whose subsistence or artistic livelihood is surplus to capitalist economies.[59] Such spaces, unrecognized as brown commons, are often erased by planners building paved paths for joggers with strollers. Sometimes these illicit uses come back, when city maintenance declines, as in the many parks along the Duwamish and the San Francisco estuary where unhoused people live. Perhaps other spaces and uses can be designed into shoreline repair projects—kids' dangerous playgrounds that leave trees and raw materials for shaping and reshaping, bathrooms and showers for the houseless, and places to have bonfires. Brown commons can often persist if cities tolerate temporary structures like those that housed and homeless locals build with driftwood on Santa Cruz beaches every winter and that city bulldozers sweep away before the summer tourists arrive. The brown commons refuses a cut along lines of respectability or class, and its demands for places to turn tricks, cruise, or shoot up are not assimilable into official planning processes.

Brown Commons Are Not Made; We Attune to Them

Now we come close to the end, having gone cruising with Muñoz through a flickering montage of urban waterfront commons, on brownfields that are either dismissed from capital's ledger book or slated for green gentrification. Muñoz is with my poet-lover and me as we scale a split rail fence under the West Seattle Bridge, back to the tiny wetland restoration site. A plaque, almost hidden by a vigorous willow, commemorates Bernice White, a Duwamish tribal member who led early Superfund cleanup actions. Little squares of plastic fencing protect plots of pickleweed from browsing geese. We walk along a cobbled slip of beach under the piers of the Seafarers Rest Mission. July and I are thinking of Gloria Anzaldúa beachcombing, of Karen Barad talking about their shared walk, as we watched ripples along the Santa Cruz breakwater, of Vanessa Agard-Jones describing what the sands remember,[60] of all the brilliant theory that queers and trans people have made while walking along the shore.

The tide is out. Muñoz stands on a broken-off piling. He's clutching his papers in his right hand, gesturing expansively with his left, then he looks up through his glasses at the old silo across the river, *LABORATORY* written on it vertically in giant airbrush letters, and says, "Brown commons are not made; we attune to them."[61] A barge comes by upriver. Two guys in life jackets sit on the side with their legs hanging off. Someone up in the tower pushes a button, and the concrete bridge swings open on a hidden hinge. The barge wake rushes up around our feet, then sinks into the sand, where cobbles imbricate and in time may become conglomerate rock. The cobbles are not rocks but concrete chunks, or pieces of brick encrusted with barnacles, or dully gleaming rusted slag with nothing growing on it. One concrete chunk once surrounded a ring of steel; this rusted hole becomes an apparatus for attuning to shoreline particulars: a frame that is in focus while the background blurs.

So now you are here with Muñoz and July and me, along the waterfront, some urban waterfront that may or may not have salmon, or small birdhouses made by local kids from painted gourds, or restoration narratives centering Boeing's corporate responsibility, but certainly has graffiti, and piers, and Brown, Black, and Indigenous people, queer and not, fishing, cruising, or sitting alone. Has this waterfront shaped you? Do you, perhaps, have a hand in shaping its future course and trajectory? With this rusty apparatus, this more-than-human prosthesis to help us see and feel the waterfront, let us together think about the politics that become possible as we attune to this place as a brown commons. Who we are matters—for the brown commons marks a kind of joy and freedom and, indeed, a genre of the human, that rejects affects of whiteness and settler fantasies of hegemony.[62]

What we see through the apparatus: The Duwamish River's banks are a mosaic of decay and contamination, rebuilt parks and the Duwamish Tribe longhouse, office parks and luxury marinas, concrete plants and dry docks. The Duwamish River Cleanup Coalition has demanded parks, trails, green buffers along freeways, and jobs in toxic remediation and salmon habitat restoration. When we center subsistence fishers and local residents, largely nonwhite and low income and about half foreign born, these groups unsettle the whiteness and settler-colonial mind-sets within Superfund planning processes. As these coalitions work to rework the river's banks and uplands, the brown commons also asks what it would mean for a place like this, which in some senses was made and remade by human hands and industry, to be unmade, or for making to include more-than-human action. What would happen if a boisterous enough multitude, including the houseless, sex workers, queers cruising, and others left out of official processes, said (in the performative sense) that brown commons are not made? What might this mean for future endeavors here, including activities like dredging, building, or placing logs in the shallows to revive salmon lifeworlds? Brown commons points to an ongoingness, and all-alongness that manifests as hairline cracks in official planning processes. It invokes the repurposing of gentrified waterfronts by people their landscaping was designed to exclude.

A Manifesto for Making Brownfields into Brown Commons

Brown commoners reject purity discourse and the fear it engenders—of toxins, of a migrant other, of a queer or trans person filthy to the core.

From brown commons comes a refusal of purity discourses in projects of green gentrification, the new urban renewal, that sweeps houseless people from along the riverbanks; ousts queers and freaks from marginal, illicit housing in postindustrial spaces; displaces punk teahouses and Lao fishing spots with smooth jogging trails and fences along the water.

Strategies of resistance developed in the brownfield waterfront commons come from desire and make a diversity of tactics for life and vitality amid rising tides. Oppositional movements shut down power plants and reject their repurposing as desalination plants. Adaptive movements repurpose "brownfield" sites as grassroots factories for the Green New Deal. Marginal movements live, fish, make art, and party in these liminal, amphibious spaces, where floods and king tides increasingly hold sway. All these strategies and more reject anthropocene logics of control and austerity. Without glorifying homelessness or informality, without relishing the tang of PCBs in a salmon filet, let us also recognize that survival strategies forged in the brown commons—making do, contact in cross-generational, mixed-race, queer encounters in the swerve—have long opposed capital's fixing and armoring logics.

Brown commoners seize on and demand queer excess of alliances made through contact. Brown commons comprise fishers and salmon on the Duwamish and might at times extend to the university and Corps of Engineers scientists—if they center the Brown (and Black and Indigenous) understanding that the salmon's recovery in the lower river is crucial to collective survival. However, whether these scientists are invited and enter into the commons—or even notice that these commons exist— depends on how and to what ends they do science. To return to Tuck and Yang and desire as a refusal of settler-colonial logics, such scientists must ally their studies with local struggles for the land and water. Interventions by outside "experts" that attempt to save disadvantaged communities from contaminated fish and crabs by limiting their catch perpetuates white affects and settler-colonial methods. Co-research must center local and Indigenous knowledge-holders' concerns and generate research methods relationally.

"Brownness is a kind of uncanny persistence in the face of distressed conditions of possibility."[63] The brown commons includes salmon that swim down and back up the Duwamish, making local and specific relations with subsistence fishers, as they do with Muckleshoot Tribe fishers with a recognized treaty right to catch them and with unrecognized Duwamish Tribe members who also fish for them. The brown commons lets us explore how such world-making relations unsettle river restoration practice, which becomes limited by property regimes to public or privatized lands or responsive only to property owners or powerful constituents. The brown commons insists on environmental justice movements' demand that affected communities lead toxic cleanups. The brown commons situates community organizing within the more-than-human, demanding that fish, birds, insects, and plants recover and thrive, and that sounds from generator shows and boomboxes persist, and that these species and sounds engender feelings of connection or solitude or thrill or grief in people who cruise the waterfront. It exalts in sanctioned and illicit encounters that happen there, in these common urban places where urban land and water meet.

Conclusion: The Allure That Permits Us to Realize That One Is Not Starting Anything

As I draft this manifesto, I imagine Muñoz reading his paper (which reads in some parts like queer-of-color and feminist manifestos) not at an academic conference but at a dry public meeting in some badly ventilated fluorescent-lit hall of power, where bureaucrats from state entities and NGOs hash out decades-long plans to repair a small fraction of the harm done to the Duwamish River. He's leaning on the podium, intent on his loose-leaf talk, skipping over parts of the argument to cut to the chase at the end of his five minutes. With some pages upside down, he finds his place, looks up at the crowd, pulls out a flag that says "critical utopianism," and waves it in the air. "It was Leon Trotsky who famously said, 'active indignation is linked with hope,'" he reads. "This hope is what I've called an educated hope or a critical desire. It's an affective refusal and salient demand for something else. Critical utopianism is not born of complacency, of an idle wishing for things to get better. It's born of the sense of indignation one feels at the harm that is visited upon humans, individuals, cultures, ways of life, the planet itself."[64] Muñoz gestures expansively with a fist full of papers, turns to the river, and exclaims, "Let the mysteries between us not function as barriers to realizing that commonness, but instead to know that as the allure that permits us to realize that one is not starting anything, but instead is fortunate enough to be a participant in something vaster, something common."[65]

The activists have their fists in the air, and they get to their feet and fall in behind him as he heads for the door, where water is starting to pour in. Outside, the air screams with gulls.

Underflow 6

Field Writing with the Brown Commons

List: Duwamish River Singularities

Riprap, creosote pilings, concrete pilings, tires as riprap and jetsam, poured concrete walls and ramps, broken concrete, slag, metal sediment (washers, screws, bits of copper, bits of wire, rusted chains), rope, line, tar globs, sharp glass, melted glass, soft sea glass, oil in sand, fuel slick glistening on water, stakes, netting and other metal wire to protect restoration sites, logs with root wads, logs placed vertically, anchor boulders, bridges, cranes, pipes and manifolds, sewer outfalls, pylons, docks, piers, fueling terminals, shipping containers, factories, train tracks, piles of aggregate, junkyards, strange sterile industrial parks, pleasure boat docks, towering hedges of blackberry, sunken boats, marinas, footbridges, parks, beaches.

Apparatus Description

A hole in a thing lets you look through it. Some holes are bigger than others, and what you look through, you see. Also, through it, you see something that is in front of you. To see more clearly, you can see through a camera through the hole. The camera, like the obscuring matter around the hole, allows you to see the depth of the field, to see the lack of focus that the human eye hides, to highlight a matter of movement or stillness from the larger field via a frame. A photograph through a hole is a frame within a frame. Focus is a trope of perfection. In landscape photography, we can think of Group f/64—Ansel Adams and his cohort—going out into the Sierras with their eight-by-ten cameras and glass plates, with their unnamed assistants and porters and mules, waiting there to capture the crisp perfection of that natural sublime as an etching of light on silver, an amalgam of mines and quarries and factories and pastures that brought them together there in the apparatus of the photograph, and all of the apparatuses around its subsequent reproduction of an image of abstract perfection that obscures forest management decisions like fire suppression, which themselves were rooted in federal policies of denying or not honoring treaties with thousands of California tribes, because they weren't as warlike as the Lakota or Modocs, because Manifest Destiny had consolidated itself enough, had settled enough, to be cavalier in its denial of sovereignty and more-than-human relation. A photograph with a hole in it mars this perfection. In color, only partially showing something that had caught my eye, the hole occludes that sense of certitude and ability to know and render as data and policy that those first photographs of the white male heteronormative environmental movement embody, present, illustrate.

A Manifesto for Making Photographs Through Holes

1. We human-camera cyborgs honor the wood borers, the rock borers, the snails and worms and burrowers in concrete, the chemical reactions that make holes in metal discarded along the shore, also the drills and human wielders of drills, taps, presses, and forges that render metal holey. These are all in the we that make photographs through holes.
2. We seek out sites that others have rendered as flat, knowable, and perfected, in their performance of industry or protected nature or new urbanism or multicultural melting pot. These sites are rendered, in official photographs as in drawings, schema, blueprints, and plans, polished and finished products of either nature or culture. Through holes, haphazardly, and with uncertain focus, we reject this fixing of views.

3. We reject official viewpoints, scenic vistas, sites that are set up to be framed and photographed by endless visitors, all the same way.
4. We want to see through more than one hole. Seeing multiply, even as it makes focus through a cell phone camera nigh impossible, invites us into the multiplicity of the world and its many eyes.
5. Holes occur everywhere; they arise from collaborations among singularities resolutely in place or in passage and have come down to us through the millennia to give us ways to see that place. Thus we reject the idea that holes can come from one place and be used to photograph another.
6. We demand utopia, the not-yet-here that holes let us see or highlight or that bring beings into our attention (if not focus).
7. When we take photographs through holes, we punch holes in the idea of the perfection of the scene.

Riddled with Holes

What is riddled with holes was once a teeming commons   worms on some old pier piling   mud flats pierced by creosote wells   shells with holes in them from other shelled singularities   pierced with a drill by a sister and worn on a neck   a neck in cross section is like a riddled pier, veins, arteries, bones, cords, holes for nerves   holes for holes   cut across a pier when you remove it   saw it up on the beach like a tree   float it from the island on a boat   wrapped in chains it is a toxic present   holes fill with water, then clams   who make their own holes in the sand   but the toxic present is still there, unless it falls off back into the water   years later after rasping on rocks or sand   it washes up in Oregon   basalt made a hole through sandstone and came flowing out   hissing and popping into the ocean   vesicular because of gases   you sit on these clinky stones   holes on holes on holes   through holes in driftwood your eye   your sunchafed cheek   glint of green windows

Epilogue

Upwelling

> What is it that makes solidarity such an elusive if not difficult practice? ... In what ways can and do marginalized subjects and communities work across their micro-specificities to align more effectively against macro-structural barriers to freedom and self-determination? What is the composition of these macro-structures of exploitation and domination and what sorts of ideological attachments do they produce to blur them from view and thus block our ability to work collectively against them? Are these structures reducible to capital, white supremacy, anti-Blackness, heterosexual and cis-male dominance, and/or the violence of the state, or is our collective unfreedom overdetermined by all of these at once and in complex ways?
>
> —GLEN COULTHARD AND LEANNE BETASAMOSAKE SIMPSON, "Grounded Normativity/Place-Based Solidarity"

How might we develop Coulthard and Simpson's fruitful concept of "grounded normativity"—the ethical framework "informed by what land as a mode of reciprocal relationship (which is itself informed by place-based practices and associated forms of knowledge) ought to teach us about living our lives in relation to one another and our surroundings in a respectful, nondominating and nonexploitative way"? What would it be, then, to think and work for a grounded relationality, at once addressed to Black placemaking, geographies, and other racialized diasporas, as well as to proprietary violences incommensurate to yet not altogether separate from Indigenous land and sovereignty? What would it mean to consider the land itself as a site of an agentive fungibility that has been conscripted into the proprietary spatialities of colonial possessiveness and constrained into geographies of exploitation that no longer serve the relationalities of presence and care that have

> for so long been its domain as a common for all? The loss of land is not just a loss of property, territoriality, power, nation, or sovereignty; it is the loss of those philosophies that derive from the relationships the land itself activates, fosters, and nourishes.
>
> —JODI A. BYRD ET AL., "Predatory Value: Economies of Dispossession and Disturbed Relationalities"

Queer is not yet here. José Muñoz opens *Cruising Utopia* with that sentence, and it stayed with me through the meandering process that birthed *Underflows*. Trans is not always visible under dirty green fishing waders, unless it is singing before 150 scientists in a darkened conference ballroom, alongside a dapper-jacketed crocodile and a tall osprey. Queer and trans feeling once roused scientists to rise to their feet to sing a libretto written by a lover of mine. There, at the Bay-Delta Science Conference, dressed in my field gear as one of an Army Chorus of Engineers, I led them in singing that our white settler dreams are forged on captive streams. Together, we felt Manifest Destiny affects as vibration in our chests and gesture traversing the hall, then sang the Riparian Dispossessed Choir's invitation to transfiguring, to the tune of the finale of *Les Misérables*:

> Do you hear the ocean's roar?
> Hear the waves beat on the shoal?
> It is the music of the elements no human can control.
> When the leaping of the trout
> draws down the thunder from the skies,
> now you must work your destiny out as the waters rise!

This musical intervention was a queer ecopoetic act. Queer ecopoetics methods, integrating story, art, and performance with scientific measurements and data, can work like a log that racks up more floating wood into a jam, then reroutes the stream of science across levees that constrain it from the world of collective feeling and grassroots politics. Most academic and settler scientists see art and the humanities as tools for communicating science, rather than as methods for transfiguring scientific thought. This may be changing, however; ecology conferences have increasingly featured artist panels or film screenings, which make a welcome respite from the policy talks. Even so, queer and trans art rarely features. For most river scientists, the idea that queer, trans, or feminist thought could make their science better, much less galvanize environmental action, remains hidden beneath the riverbed of mainstream environmental governance. To assert queer performance, as well as art, maps, and storytelling, as integral to river science is to carve away these norms like a river that slowly undermines its cut bank.

If to queer is to make strange, then I hope that *Underflows* has made strange the world of river governance. While expert scientists from settler government agencies and water utilities often act as though their decisions determine what happens on the ground, this placid surface flow through conference rooms is only part of the story. Water conflicts roil like currents that scour the depths and banks of official policy; underflows well up, contest, and change which methods and worldviews matter in river management. Whereas official river policy formulates water problems as human-environmental conflicts, an underflows approach rejects this formula as both too narrow and too vague. The framing of "human-environment" is too narrow because it leaves out politics and cultural values that people make in places, with historically contingent knowledge. This framing is too vague because it collapses divergent settler, Native, *arrivánt*, and immigrant concerns into an undifferentiated human. These limitations of vision recapitulate the Manifest Destiny thinking that spawned settler-colonial water projects, by continuing to erase Native sovereignty demonstrated through ongoing Native management, and by creating situations that pit Black, immigrant, and Native people against one another. Kyle Powys Whyte's term "industrial settler colonialism" is useful in understanding this ongoing dispossession as inscribed in the land, in institutions, and in governance:

> One should not underestimate the physicality and scale of the US settler homeland creation process. Europeans, and eventually US Americans, had to physically shape the lands and waters to reflect their future aspirations and fears, economic systems, cultures, ways of life, and heritages. They literally had to carve out, or inscribe, a homeland for themselves, within a territory whose ecosystems were already coupled with the social, political, and cultural institutions of different

populations (e.g., the Anishinaabe seasonal round system).... Lacking long-term knowledge of the ecosystem and applying their own understandings from elsewhere to the land, they established large-scale industrial agriculture, factories, transportation systems, and hydrotechnologies such as dams. These technological systems replaced indigenous technological systems, altering the ecosystems through clearing land, pollution, and the construction of barriers and arteries (e.g., roads, pipelines, and dams). At the same time, industrial processes gradually distanced many settlers from an awareness of how their societies were based on these forms of natural, animal, and human resource exploitation.[1]

Underflows' proposition is that river scientists and managers should transfigure such industrial systems of water, river, and salmon management to foreground underflows of knowledge. I have argued that scholars in feminist STS and critical physical geography (CPG) can deepen their analyses of power and exclusion in environmental management by responding to Indigenous studies and allied critiques. By working across these fields and integrating queer and trans praxis, *Underflows* has traced affinities among patterns of thought and modes of collaboration, which have been forged through antidam and salmon recovery struggles by river workers, activists, fishers, humanists, and social scientists. Working from this thought will be especially crucial for a CPG and STS that engage environmental justice and Indigenous movements. These movements make like underflows and well up into river governance's surface flows to transfigure river restoration processes in order to center Native protocol and Black liberation. Queer/Two-Spirit writers and activists are central to these movements and produce some of the strongest writings on relation to the land. To do science from the queer and trans politics I have described in this book, it is necessary, but not sufficient, to understand landscape degradation and ecological processes that dominate now as coconstituted with settler colonialism. We scientists must also take up queer Indigenous and queer-of-color critique as a strategy for tracing affinities and solidarities and for transfiguring environmental governance to center relationality and place-based communities' priorities. As Jodi Byrd, Alyosha Goldstein, Jodi Melamed, and Chandan Reddy argue, "Because they have long understood inclusion in the liberal capitalist state as an unmaking of collective social being, as whitewashing ongoing violence, Black radical philosophy read alongside Indigenous critical theory may help us to think alternatives that are both transformative and realizable and, in fact, already manifest and discernable."[2]

Underflows' project has been to bring these conversations into the realm of environmental studies. My ecology studies of salmon's life chances under harsh conditions that become more common as climate change accelerates, and how they fare in new habitats that humans or beavers construct, began from a desire to find alternatives to Manifest Destiny river governance. For me, understanding of river drying and salmon extinction as inextricable from settler-colonial policies developed gradually, through study of Indigenous scholarship and in conversation with Native collaborators.

Trans slaps the placid water of science as usual, in the guise of a beaver dragging something across its new pond. Small fish flop in the recovery bucket that I carry back to the stream, then tip, releasing them to swim as they may in shrinking stream pools. To draw together this book's threads of queer and trans feeling into a vision for doing science differently, let me take you down, once more, to a river reach, the Mid-Klamath, one where I am making the queer more explicit in ecology and bringing together Indigenous and trans ways of feeling. In sketching some of the practices and relations I and collaborators are making, I revisit the questions, What are the intersections between being queer and doing queer ecology? How does a community of queer ecologists help support queer ecological thought? How can cis or straight people enter into queer trans ecology practices? I frame these questions thus because of conversations I've had with Indigenous scholars and practitioners about how settlers can work in solidarity with Native and Indigenous studies and Native-led research projects. The specific answers to questions of Native–non-Native solidarity are always place based and relational; we, those undertaking a project, create them through protocol.

Alongside the theoretical work collected in *Underflows*, I have begun new collaborative river science projects, which I place, explicitly, at the nexus of queer, trans, and Indigenous thought. One new project, coconceived with Karuk basket weaver and educator Lisa Morehead-Hillman and cultural practitioner and ceremonial leader Leaf Hillman, works toward grounded relationality at Tishánik, a village and ceremonial site on a floodplain blasted and degraded by mining. More broadly, we hope to transfigure river governance on the Klamath River to center Karuk and other Indigenous protocols and priorities. In our conversations in the field and around the fire, and in our early collaborative writing, queer and trans thought meets and responds to Karuk understandings of river relations and how to restore them. From this joint thought, we are developing a proposition that underflows are floodplains' memories. We mean this in two senses—that subsurface flow paths of water through sediment and muck hold the river forms that sustained ecocultural relations before mining and that Karuk knowledge, activated by caring for the land, carries crucial information on the shared harm and collective powers of resurgence. In the rest of this epilogue, I situate this new project within a longer trajectory of Klamath River governance, examine its queer and trans potentials, then propose underflows strategies that could travel to other rivers and lands.

Tishánik remains a site of cultural connection and salmon, manzanita, and willow persistence. Mining rerouted the river across the valley, where it erodes the hillslope under a major access road. Because tailings block most floods, there is little habitat for fish and basketry plants. Leaf Hillman and other Karuk leaders have tried for decades to mobilize state, federal, and local agencies to move the river back into the floodplain. Lisa Morehead-Hillman invited me to collaborate on a plan, centering our work in intergenerational knowledge exchange among Karuk youths, elders, and cultural practitioners. We have begun training Karuk student interns in interviewing and mapmaking; in summer field courses, they are documenting the river, its history, and its ongoing cultural connections and writing and mapping their dreams for its future. The students interviewed Leaf and Lisa in the shade of locust trees that surround the ceremonial camp that is inhabited each August during pikyávish, the world renewal ceremony. Miners planted the locust trees after they blasted away the hill and riverbank all around the village. Trees have taken over here, while blackberries and star thistle sprawl across the tailings below. Leaf and Lisa tell us this history: miners killed Karuk men and raped Karuk women, and they worked to evict those who escaped by calling them squatters. This gendered violence was and remains crucial to dispossession, in a process similar to what Coulthard and Simpson describe in Canada: "The state-sanctioned murdering, assimilating, and disappearing of Indigenous bodies (asymmetrically distributed across genders) are, as the Mohawk scholar Audra Simpson says, a direct attack on Indigenous political orders because these bodies generate knowledge, political systems, and ways of being that contest the hegemony of settler governmentality and thus make dispossession all the more difficult to achieve."[3]

At Tishánik in July 2020, five Karuk high school kids and I ride out in a pickup bed, bouncing down the pitted tailings road to a manzanita grove high above the river. Kathy McCovey, a Karuk cultural practitioner and archaeologist, shows the students how Indian potatoes grow between the manzanita bushes; both species are thriving in the harsh heat that bakes the head-sized rocks. We sample manzanita berries—chalky sour-sweet surrounding rock-hard seeds—as Kathy asks the students to look for scat. They find no fresh bear scat; Kathy thinks that's because the manzanitas are fruiting two weeks earlier than usual. She is one of the core researchers in a study of climate change influences on cultural focal species and is working on understanding how bears and Karuk people can adapt. Kathy says the river once curved toward the village and describes the pikyávish dances held on its banks. She remembers people describing a deep fishing hole and salmon hideout at the Camp Creek confluence, when that confluence was next to the village.

These talks at the river are part of Karuk protocol. Though my informal relations with Karuk natural resources staff began in graduate school, my first formal engagement with the Tribe was in 2017, when I proposed a University of Washington field course to the river. Lisa, then head of the Tribe's Píkyav Field Institute, proposed that Karuk high school and college-aged students participate in the class and earn college credit. I worked the university bureaucracies to make that happen and hustled funding to bring the Karuk students to Seattle for a five-day data analysis workshop.

Lisa's invitation and my response laid the groundwork for our ongoing collaboration, one that centers Karuk science and community benefit and also integrates queer and trans protocol. During the 2018 course, students and faculty met for breakfast in a Forest Service campground that occupies a Karuk acorn camp. On the last day, Lisa and July Hazard and I talked while our two-year-old kid, Blue Jay, brought us flowers and twirled in his favorite dress. Our conversation ranged across geography, poetry, rivers, theater, and language. Lisa mentioned her husband Leaf Hillman's work on mining histories and his long-standing desire to study how to realign the river to a regenerative path. She invited us to pikyávish at Tishánik, explaining that in Karuk protocol that was the proper time and place to discuss work there with Leaf, the ceremonial leader responsible for the Orleans area. Engaging in queer protocol, Lisa let us know that we were all welcome at the ceremony—"you and your son, katch katch" (Karuk for blue jay). Engaging in Karuk protocol, we thanked her for the invitation. I said, "We call him our child; he'll let us know what his gender is later." That exchange began several years of conversation on queer, trans, and Karuk strategy and belonging. In long van rides up and down the river during our 2018 class, Karuk and UW students discussed settler masculinity's influence on Karuk and settler Seattle culture. Later, Lisa and Leaf told me and my PhD student Sofi Courtney (who is nonbinary) that queer young people are coming out and into more welcome and acceptance. Lately, Hazard and I documented some of these conversations in a podcast and made and plans for a queer ecologies reading group on the river.[4]

When we discuss our collaboration, Lisa and I have different ways of describing our blend of Indigenous and queer ecologies. We assert that these theories and practices need one another and can make each other stronger. Our collaboration is slowly building a shared space of encounter, for the queer and trans students who are gravitating to my lab and for Karuk students navigating complex gender expectations while seeking livelihoods that will let them work close to home.

Within Native-led collaborations like the one at Tishánik, grounding science in storytelling and knowledge sharing while in the field together is standard practice. Lisa, Leaf, and I call this practice Karuk science. In my collaborations with other settlers, doing ethnography while doing science is a queer practice in that it perverts the disciplinary norms of both disciplines. It reveals science's sociality, and it reintegrates politics and policy into science question-making and analysis. These practices expand science's reach. Strategically, grounding ethnography in collaborative science allows me to support collaborators' political struggles for land and governance authority, by providing field equipment, labor, and analysis expertise that produces scientific data that they can use in their legal cases or policy work; I also raise funds to directly support Karuk Tribe programs.

At Tishánik in fall 2020 with Lisa, ecologist Frank Lake (Karuk descendant), and three non-Native collaborators, this approach yielded a research design for plant responses to flood and fire. In their conversation, Lisa and Frank switched effortlessly between story, number, and ecocultural practice. They discussed statistical methods to quantify how basket weavers' harvest and burning affects willows, hydrological cycles of flood that scour willow roots and sprout new cottonwoods, and stories of moving rocks to harvest Indian potatoes up on the terrace. Their core concern was how Tishánik could become—indeed was becoming—a classroom and community gathering place. Already, dozens of Karuk basket weavers and gatherers are stewarding the hardy plants that are thriving in the harsh tailings piles. Next, we could remove some tailings and redirect the river to create pools, sandbars, and slow-moving side channels that can support species that once thrived there. The non-Natives—me, Sofi Courtney, Shawn Bourque, and my longtime collaborator Dan Sarna-Wojcicki—filled out phenology surveys for the tribal climate science project and shared ideas on how to visualize and analyze options for reconfiguring the river channel. Sciences flowed back and forth through the conversation, interwoven with threads about miners' gender-based violence, how dances persisted at Tishánik as miners sluiced the river bar away, and Lake's career-long struggles to normalize cultural burning within Forest Service management plans.

Such meandering conversation happens in settler collaborations, too, and most good field studies originate in this kind of cross-pollination. What felt different to Sofi, Shawn, Dan, and me was that centering the science on cultural practitioners' knowledge gave the science a direction, an ethical orientation to care for tribal members and intergenerational relations. The approach we're weaving at Tishánik draws on the most radical and reflexive tendencies of feminist, queer, and trans thought—Indigenous, postcolonial, and queer-of-color theory—that emphasize accountability and reciprocity. These modes and methods activate latent potentials in riverine sciences—to act from an antiracist and unsettling politics to shape agency policy, and to make science with a broader collective, building trust and solidarity for the long haul: work to revive flood, fire, and the migrations and regenerations they renew.

Through such conversations, we settlers can see rivers in a new way. A river drawn as a single blue line on a map is at best a two-dimensional snapshot of constantly shifting four-dimensional movement and transformation of energy. Floodplains become slow-moving, shallow lakes during floods, acting as sponges that release water to seep into deep aquifers or to well up into river channels or wetlands. As property boundaries, surface channels—that blue line on a map—are fickle. They jump their banks or bifurcate, in a process geomorphologists call *avulsion*, meaning a tearing away. In flood, floodplains teem with life, as algae and insects proliferate in shallow, nutrient-rich waters and become food for fishes, birds, and their predators and decomposers. In the floodplain, water plays one of its most dynamic and vital roles, burying and excavating land in an ever-shifting mosaic that favors different plant communities in turn. In some sense, the floodplain itself *is* the river, in turn wet and dry on the surface but sustained underneath by constant hyporheic flow.

If the blue-line river is a settler desire for rivers to remain fixed between confining levees, then underflows are floodplains'. memories.[5] They seep through gravel of old point bars or sand that settled out in eddies. They stagnate in lenses of black earth made from decomposed plant parts. As floodwaters percolate down into floodplains, they flow more quickly through gravel and more slowly through sand; clay stops them. These sediments mark old river processes—the gravel dropped out on top of a river bar in a flood, the sand settled out in a backwater eddy, and the clay drifted down slowly on an older floodplain. Moving through these substrates, underflows transport carbon and nutrients that drive subsurface microbial respiration. They carry acids, metals, and organic toxins, and they precipitate them in new forms.

The streams that salmon rear in flow through timber and farmland and timberland being carved into residential lots and vineyards and crisscrossed with roads. Responding to these settler cuts, streams eroded down into the floodplains to flow through narrow trenches; water tables dropped apace. Decolonizing means returning land and management authority and resources to Native nations, many of which prioritize acquiring riverine lands and reconnecting floodplains to rivers.[6] Streams in summer-dry climates like California need to flood in winter if they are to flow in summer, need to back up behind logjams and spill across the land. Salmon need these logs and waters to make still hideouts where they can escape torrential flows, which scour deep pools, then recede. Floodwaters sink into the ground as salmon move into the pools, which stay cool in summer when underflows seep back into the stream. For floodplains to flood, some settler agriculture and residential land will need to flood again, and this action of flooding would begin to reshape the physical colonization that Whyte describes.

Settlers long tried to stop floods. Sometimes, for a while, they succeeded in routing floodwaters away from infrastructure and agricultural lands. They changed the land so water ran off it, severed streams from floodplains with levees, suppressed fires, and trapped beavers that kept riparian forests open. As settlers continue to maintain physical infrastructures and extract wealth from the land, they build new environmental policies on old legal and regulatory orders, continuing dispossession.[7] But the idea that settlers can control rivers without cost is a fallacy; the costs are brutally clear in dried-up rivers, wildfires, poisoned lands, and extinctions. Only by restoring governance authority, land, and collective capacity to Native nations will this ecological and cultural rift begin to suture. Recognizing this, some settler residents, in the Klamath and elsewhere, are working to unsettle through everyday actions of tending the land and by taking political action to block pipelines and bring down dams.

River Justice on the Mid-Klamath

On the Klamath, dam removal struggle began two decades back, when a tribal coalition and settler allies challenged the federal relicensing of four salmon-blocking dams. One could trace this struggle through its surface flows—Federal Energy Regulatory Committee and National Marine Fisheries Service hearings, the dam owners' shareholder meetings,[8] lawsuits brought by Native nations and fishermen's associations, congressional hearings, and press conferences featuring tribal leaders, governors, and agency scientists. Yet to do so would miss critical underflows that pushed or redirected those official processes. At every slowdown, impasse, or failed deal, water protectors took direct action. They painted banners, staged boat blockades, sang and held ceremonies, cooked and shared salmon, testified, wrote reports and press releases, and worked public opinion channels. They moved as underflows that well up into the surface, then seep back down the hyporheic to recharge deep aquifers. Tribal leaders shifted the stage of debate by bringing leaders of fisheries agencies and governors down to the river to catch and eat salmon, and to see salmon dying from parasites and toxic algae that thrive in the stagnant reservoirs. Allies reached out to CEOs and shareholders through family networks or raised money or visited the dams and posted to social media with our own homemade signs. We all continue to do this relentlessly, like salmon swimming upstream or beavers repairing a breached dam, and we forge a collective life in the struggle and telling and retelling of the river's stories. As I write this in mid-2021, these efforts have again breached a regulatory impasse, and the dams are on schedule to come down in 2023. We remain on alert but are also starting to plan the celebration for when the salmon surge up into the rivers they've dreamed of for so long.

Queer Ellipses, Trans Potentials

In the Tishánik project, we collaborators are using ecopoetics strategies that make sense within Indigenous and trans collective practice, working first with Karuk elders and cultural practitioners to develop a vision for the river's future. Once we've integrated stories, maps, and ecological studies into a conceptual model of justice for the river, we will go public with these results, by convening Native and settler residents, scientists, and policy makers for a workshop. We hope that this event will shake up river governance processes that have become stuck in conflict and agency neglect. This public event will invite participants into Karuk protocol, in order to center scientific controversies and engineering challenges in Karuk goals, knowledge, and priorities. Where queer and trans strategies sync with Karuk protocol, we will mobilize them to animate affective connections and to probe collective feelings for the river and its future.

Queer performance can produce new thought and analysis that changes the guiding questions and enabling conditions for river science. Performance, a key queer trans aesthetic strategy, once led me and Daniel Sarna-Wojcicki to theorize beavers' ecological effects with a boundary-transgressing theory we called "the amphibious":

> Although salmon, unlike beavers, cross over into the terrestrial realm only in death, to consider them amphibious in this broader sense that we are elaborating does useful work, we think, in embodying material

connectivities among living organisms and their constituent elements and relations in floodplain ecosystems. The amphibious can bring to light a dynamic, relational ecology in which humans' and other bodies are, only briefly, containers for the elements that circulate endlessly through the landscape and waterscape. Against instrumentalist appropriations of beaver dams to engineer better rivers and/or society, we see in beaver dams/ponds rich multispecies cascades of living and dying that comprise the hyporheic zone and the potential to guide collaborative watershed restoration.[9]

This insight grew out of two performances Sarna-Wojcicki and I staged in 2013 and 2014, which unsettled imaginaries of settler river control by advocating transspecies collaboration with beavers.[10] The science that we are now undertaking grows from this theory and the proposition that beavers might be transfiguring agents working to overturn Manifest Destiny.[11] In the performance, he intoned, "Part of the project of Manifest Reversal entails viewing the actual process of manifest conquest as incomplete, fragmentary, and subject to multiple forms of resistance. Even as fur brigades snagged millions of pelts and wiped out beaver colonies across the US, there were instances of microresistance by both humans and beaver themselves." Just as beavers had disrupted human restoration plans by stealing their willows, they chewed into our collaborative research project, inspiring our transgressions of standard academic forms and our undisciplined conversations. Such queer performance strategies and trans relational thought can produce new collective affects of outrage and love, both within the people who stage the performances and within the audiences who witness them. The very idea of performance, of being together conspiring, takes on a dreamlike, almost impossible quality as I write this in the depths of the pandemic. Yet it may be that when we can finally breathe together in person again, such strategies and affects will hit all the more forcefully, opening up new radical politics.

If queer, trans, feminist, and Indigenous approaches work together, along the Klamath and elsewhere, what changes for human researchers and the riverine community? First, these approaches enable researchers to consider justice relationally by acting to materialize the feminist, queer-of-color, and trans senses of becoming-with.[12] Within this framework, it makes sense to talk of a river's "looking good," in the Karuk sense of peeshkêesh yáv umúsaheesh ("the river will look good," meaning that it will make a good pattern of flow that sustains cultural practice). Queer theorization of failure might let us see messy avulsions (when a flooding river moves dynamically and flouts settler levees and dredge tailings' confines) as so bad that they are good, because disturbance creates conditions of possibility for new configurations of human and more-than-human relation. Queer kin making happens in places like Tishánik in a similar way that it does along the Duwamish, through care practices that refute Western science's definition of such landscape as irrevocably damaged and disposable. Trans embodiment and affinity for ongoing change insists on abolition of white supremacist thought and structure as part of rehabilitating damaged landscapes and multispecies relations.

Approaching the Same Stream Again and Again

The stakes of bringing queerness into ecology via aesthetic theory and interventions is an active assertion of more relational politics and values. Lisa's naming of our project peeshkêesh yáv umúsaheesh, and translating the different senses of looking good from the Karuk, triggered a cascade of queer and trans associations with "looking good." In queer culture, to look good is to look fabulous, sexy, embodied, ready for a night on the dance floor, probably involving leather, skintight shimmery fabric, glitter, and maybe sequins. Trans looking good might mean passing, but it just as well might mean embodying an in-between, androgynous, genderfuck, or fluid state. It might mean strutting your stuff with confidence and lots of makeup, whatever you've got—it can be an attitude as much as a look, and one that often challenges and subverts heterosexist notions of beauty and the erotic. The queer aesthetic is fishy pleasure—both pleasure we queers take in being and swimming with fish and pleasure in embodying fish and other river creatures in glittery costumes we make from cast-off fabric and papier-mâché. In embodying fishy pleasure queerly, performers show scientists and others how to imagine being fish, take pleasure in being with fish, and thus be able to imagine fishy belonging in a transfigured river that rejects settler confines and norms.

In the Scott Valley and Mid-Klamath, my own work toward such alliances is moving ahead. For over a decade, questions of the extent and temporality of the groundwater–surface water connection and hyporheic exchange have driven tribes and agricultural associations to commission competing hydrologic models and field studies. As I have demonstrated in this book, a feminist science studies approach can trace political and ethical commitments through governance processes. This approach can reveal how such processes, at different times and in different places, are coconstituted with Indigenous and settler worldviews and ways of knowing. In the Scott Valley, more than any other place I have worked, potential collaborators ask me, as an outsider, non-Native, university-affiliated researcher, to state my stakes, my relation to the place and its human communities, and my political and ontological commitments. Here, everyone believes that science is inherently political, and so this approach that I have outlined, of situating science within cultural and political frameworks, resonates with many collaborators. Feminist science has created a space, at least temporarily, to maintain dissensus about remedies while pursuing matters of mutual concern. In ongoing collaborations around groundwater modeling and beaver-mimicking restoration projects, the partners maintain their divergent stances vis-à-vis groundwater models and groundwater and surface water forums. While allying for the short term, they reserve the possibility of mobilizing distinct political strategies.

The surface flows of environmental science put forth the idea that once we get enough information, we'll make a change in management. Yet despite overwhelming evidence, old policies and patterns of land use persist, and so degradation and species loss accelerate. Deep transformation in settler governance of rivers, I propose, will come not from merely accumulating more facts but by communicating the emotional life of rivers and place-based communities that stake their lives on them. As long as the affective realm is devalued in decision-making, policies and institutions will continue to exclude the Native and non-Native communities that make politics guided by love for rivers. Within this broad movement, reembodiment as a collective watershed body and practicing queer reattachment to place by publicly expressing love for riverine species—these acts make possible a redrawing of relations that challenges settler power structures. As in the undercommons, this coming-together is really the mode and mechanism for change. Upwelling from the subsurface, underflows methods propose that rational argument is not the main change maker; what makes change, rather, is how we draw people in and make communities in struggle. It's not the data but the field-made relationships that enable this change.

I have tried to draw together this book's underflows, in a contingent way, and to show some ways that transfiguring rivers and queering ecology might happen and how that might matter for theory, for Native-settler solidarities, and for queer and trans being together.

Thinking about what kind of science grows out of feminist theory, Banu Subramaniam and Angie Willey asked, "What is the proper object of science?"[13] Adapting their question to the rivers I love, I begin a list.

What is the proper object of river ecology?
water
dams
fish
models
intra-actions
beavers
indeterminacy
erosion
deposition
scour
flooding
apparatuses
imaginaries
resurgence
transfiguring
melancholia
performativity
transgression
relations
futures

I hope that you—through your own water relations—will add to this list as you transfigure science into a queer relation, in all the rivers you swim in.

Notes

Prologue

1. PIT (passive integrated transponder) tags are rice-sized RFID transponders similar to microchips commonly used in pets. Their unique numeric code can be read by a handheld scanner (as while tagging fish) or by a large antenna in the water, which is made of copper wires encased in PVC pipes that extend across a stream. This data is used to track salmon through different life stages (fry, parr, smolt, and returning adult) and estimate population dynamics.
2. In water rights, the principle of beneficial use describes how water can be diverted from a stream—generally for agricultural, municipal, industrial, and residential use, so long as the diverter uses only as much as is needed. In-stream flows for fisheries have not historically been considered a beneficial use. As a result, in recent decades environmental advocates and Native nations have used the doctrine of public trust to argue that some water must remain in streams and lakes in order to preserve aquatic species, in which the public has an inherent interest. For a brief overview of California's interpretation of these principles, see California State Water Resources Control Board, "Water Rights: Public Trust Resources."
3. This section grew out of conversations with Daniel Sarna-Wojcicki and appears in an earlier form in Woelfle-Erskine and Sarna-Wojcicki, "Hyporheic Imaginaries."

Introduction

1. Muñoz, "Preface." Chapter 5 thinks deeply with Muñoz's idea of the brown commons, still largely unpublished at the time of his death, in 2013. When I began writing this chapter in 2015, and until the publication of *The Sense of Brown* in 2020, this was the longest extant writing on the

brown commons, and the text in which this quote appears is Muñoz, "The Brown Commons."
2   Taylor, *The Body Is Not an Apology*.
3   Taylor, *The Body Is Not an Apology*; Doroshow and Walker, "The Black Trans Lives Matter March Was This Year's Pride."
4   See, for example, Daigle and Ramírez, "Decolonial Geographies": "Although in many ways distinct, Indigenous and Black women, youth, queer, Two-Spirit and trans individuals continue to be subjected to interconnected forms of gendered colonial and anti-black violence in settler colonial contexts, thus bodily sovereignty is essential to liberation, particularly as it is a crucial site of knowledge and self-determination."
5   For an excellent hydrological exploration of this uncertainty and dynamism, see Wheaton, Darby, and Sear, "The Scope of Uncertainties in River Restoration."
6   For a critical discussion of environmental themes in speculative fiction, see Canavan and Robinson, *Green Planets*. For examples of fiction in this genre, see Robinson, *Antarctica*; FUTURESTATES, "The 6th World"; Adams, *Loosed upon the World*; Simpson, *This Accident of Being Lost*; Dillon, *Walking the Clouds*; McHugh, *After the Apocalypse*; Butler, *Parable of the Sower*; Robinson, *2312*; Ming-Yi, *The Man with the Compound Eyes*; Okorafor-Mbachu, *Zahrah the Windseeker*.
7   Whyte, "Our Ancestors' Dystopia Now"; Tuck and Yang, "Decolonization Is Not a Metaphor"; Ferguson, "To Catch a Light-Filled Vision."
8   Deen et al., "Thirsty for Justice."
9   Woelfle-Erskine et al., *Dam Nation*.
10 White, *The Organic Machine*.
11 Martinez, "Protected Areas, Indigenous Peoples, and the Western Idea of Nature."
12 Norgaard, "The Politics of Fire and the Social Impacts of Fire Exclusion on the Klamath."
13 Cronin and Ostergren, "Tribal Watershed Management"; Trosper, "Resilience in Pre-contact Pacific Northwest Social Ecological Systems."
14 Trosper, "Resilience in Pre-contact Pacific Northwest Social Ecological Systems."
15 I worked on a crew led by Joel Glanzberg, sometimes assisted by Louie Hena (Tesuque Pueblo) at Living Structures, Inc., a natural building company that did small-scale watershed restoration and eco-landscaping.
16 I studied with professors Johnnie Moore and Andrew Wilcox in the Geology Department at the University of Montana, while working on studies of the Milltown Dam removal.
17 I trained with Stephanie Carlson, Mary Power, Michael Fawcett, and Sierra Cantor.
18 Jasanoff and Kim, "Sociotechnical Imaginaries and National Energy Policies."

19 Nadasdy, *Hunters and Bureaucrats*; Todd, "Fish Pluralities"; Weir, *Murray River Country*; Celermajer et al., "Multispecies Justice."
20 Here I am thinking with McKittrick, *Dear Science and Other Stories*.
21 Nancy, *Being Singular Plural*.
22 Malone, "Using Critical Physical Geography to Map the Unintended Consequences of Conservation Management Programs"; Schell et al., "The Ecological and Evolutionary Consequences of Systemic Racism in Urban Environments"; Schell et al., "Recreating Wakanda by Promoting Black Excellence in Ecology and Evolution."
23 Agrawal, "Dismantling the Divide between Indigenous and Scientific Knowledge"; Aldern and Goode, "The Stories Hold Water"; Anderson, *Tending the Wild*; Diver and Higgins, "Giving Back through Collaborative Research"; Holtgren, Ogren, and Whyte, "Renewing Relatives: Nmé Stewardship in a Shared Watershed"; Kimmerer, *Braiding Sweetgrass*; Martinez, "Protected Areas, Indigenous Peoples, and the Western Idea of Nature"; Mbilinyi et al., "Indigenous Knowledge as Decision Support Tool in Rainwater Harvesting"; Todd, "Fish Pluralities"; Weir, *Murray River Country*; Whyte, "Food Sovereignty, Justice and Indigenous Peoples"; as well as numerous conversations with Native and non-Native employees of tribal fisheries and natural resource programs, in the field and at workshops and conferences.
24 Whyte, "Food Sovereignty, Justice and Indigenous Peoples," 5.
25 Byrd, *The Transit of Empire*; Coulthard and Simpson, "Grounded Normativity/Place-Based Solidarity."
26 River restoration often involves transdisciplinary groups, which include scientists from universities and tribal and settler resource conservation agencies, environmental NGOs and local residents, and agricultural, timber, or industrial landowners. Transdisciplinary management and participatory action research translate lay knowledge into Western science terms and then communicate the scientific results back to lay language.
27 Turnbull, *Masons, Tricksters and Cartographers*; Scott, *Seeing Like a State*; Deloria, *Red Earth, White Lies*; TallBear, "Standing with and Speaking as Faith."
28 Haraway, "Situated Knowledges"; Harding, "'Strong Objectivity'"; Harding, *Sciences from Below*.
29 Carrera et al., "Community Science as a Pathway for Resilience in Response to a Public Health Crisis in Flint, Michigan"; Kimmerer, *Braiding Sweetgrass*; Diver, "Giving Back through Time"; Corburn, *Street Science*; Beckett and Keeling, "Rethinking Remediation"; Great Lakes Indian Fish and Wildlife Commission, "Mercury Maps."
30 Wilson, *Research Is Ceremony*; Grossman, *Unlikely Alliances*; Daigle, "Resurging through Kishiichiwan"; Whyte, Caldwell, and Schaefer, "Indigenous Lessons about Sustainability Are Not Just for 'All Humanity.'"
31 Willette, Norgaard, and Reed, "You Got to Have Fish."

32   Balazs and Morello-Frosch, "The Three Rs"; Cashman et al., "The Power and the Promise"; Corburn, "Community Knowledge in Environmental Health Science"; Fortmann, *Participatory Research in Conservation and Rural Livelihoods*; Holifield, "Environmental Justice as Recognition and Participation in Risk Assessment"; Kimmerer, *Braiding Sweetgrass*-community-based participatory research (CBPR; Kiparsky, Milman, and Vicuña, "Climate and Water"; Kirchhoff, Carmen Lemos, and Dessai, "Actionable Knowledge for Environmental Decision Making"; Liboiron, "R-Words."

33   Woelfle-Erskine and Cole, "Transfiguring the Anthropocene."

34   Although I don't rigorously prove the resulting social-ecological ramifications, such studies of changes in habitats, management of species, and social institutions would add empirical value to the theoretical approaches I propose here.

35   Edelman, *Trans Vitalities*, 125–26.

36   Chen, *Animacies*; Haritaworn, "Decolonizing the Non/Human"; Tuck and Yang, "Decolonization Is Not a Metaphor."

37   Worster, *Rivers of Empire*.

38   Li, *The Will to Improve*.

39   White, *The Organic Machine*.

40   White, *The Organic Machine*, 4.

41   Linton, *What Is Water?*

42   Linton and Budds, "The Hydrosocial Cycle"; Reisner, *Cadillac Desert*.

43   I gained this insight through informal conversations with other queer and trans scientists but also in a workshop called Queer Ecologies as Environmental Justice Strategy, which I facilitated with Eli Wheat at UW Seattle on May 9, 2020.

44   Instead, many work for visibility within science through gatherings at conferences and social media campaigns. See 500 Queer Scientists, "500 Queer Scientists Visibility Campaign."

45   Schell et al., "The Ecological and Evolutionary Consequences of Systemic Racism in Urban Environments."

46   Woelfle-Erskine, "Fishy Pleasures."

47   Gandy, "Queer Ecology."

48   I'm thinking of literature, film, and art by rural queers and of other vibrant production—online and otherwise—by young Two-Spirit people, queers, and gender deviants who visit or work on rural land.

49   Korinek, *Prairie Fairies*; Herring, *Another Country*.

50   Morgensen, *Spaces between Us*.

51   Like majority-white spaces in cities, some queer land projects have responded to critiques by Black, Indigenous, and POC members, by changing how they are run or dedicating land and resources to BIPOC-only activities.

52   Cairns and Heckman, "Restoration Ecology."

53 Whyte, "Our Ancestors' Dystopia Now"; Todd, "Relationships—Cultural Anthropology"; Kimmerer, *Braiding Sweetgrass*; Kimmerer, "The Covenant of Reciprocity."
54 Tomblin, "The Ecological Restoration Movement."
55 Whyte, "Food Sovereignty, Justice and Indigenous Peoples."
56 Beckett and Keeling, "Rethinking Remediation"; Hanak et al., *Managing California's Water*.
57 Subramaniam and Willey, "Introduction."
58 The one key exception that I have found in fisheries science is Yoshiyama and Fisher, "Long Time Past," which describes the origin of salmon hatchery science at the Baird Hatchery on the McCloud River. The paper details Livingston Stone and the US Fish Commission's appropriation of Winnemem Wintu knowledge of salmon ecology from the Winnemem families that Stone hired to propagate Chinook salmon and steelhead.
59 Turnbull, *Masons, Tricksters and Cartographers*; Nadasdy, *Hunters and Bureaucrats*; Subramaniam, *Ghost Stories for Darwin*.
60 Delany, *The Motion of Light in Water*, sec. 61.
61 Whyte, "Food Sovereignty, Justice and Indigenous Peoples."
62 See, for example, Denham et al., "Sustaining Future Environmental Educators"; Schell et al., "The Ecological and Evolutionary Consequences of Systemic Racism in Urban Environments"; Schell et al., "Recreating Wakanda by Promoting Black Excellence in Ecology and Evolution."
63 Woelfle-Erskine, Larsen, and Carlson, "Abiotic Habitat Thresholds for Salmonid Over-Summer Survival in Intermittent Streams"; Larsen and Woelfle-Erskine, "Groundwater Is Key to Salmonid Persistence and Recruitment in Intermittent Mediterranean-Climate Streams"extended intermittency can drive high mortality as pool contraction decreases pool quality, and some pools dry completely. We evaluated the influence of a suite of abiotic habitat characteristics on the over-summer survival of two imperiled salmonid fishes (coho salmon Oncorhynchus kisutch; steelhead trout Oncorhynchus mykiss; Woelfle-Erskine, "Collaborative Approaches to Flow Restoration in Intermittent Salmon-Bearing Streams"; Woelfle-Erskine, "Rain Tanks, Springs, and Broken Pipes as Emerging Water Commons along Salmon Creek, CA, USA."
64 Daigle and Ramírez, "Decolonial Geographies." The idea of place-based constellations in theory and practice comes from Simpson, *As We Have Always Done*.
65 Muñoz, *Cruising Utopia*.
66 This apparatus is inspired by Barad, *Meeting the Universe Halfway*, which students read in excerpt.
67 TallBear, "Standing with and Speaking as Faith."; Willey, "Biopossibility"; Subramaniam, *Ghost Stories for Darwin*.
68 For an expanded discussion, see Harding, *Sciences from Below*.
69 Muñoz, "Feeling Brown, Feeling Down," 675.

70 Sedgwick, *Tendencies*.
71 Hayward, "Lessons from a Starfish"; Woelfle-Erskine and Cole, "Transfiguring the Anthropocene."
72 Harney and Moten, *The Undercommons*.
73 Freeman, "Queer Belongings."
74 Muñoz et al., "Theorizing Queer Inhumanisms," 210.
75 Woelfle-Erskine, "Who Needs Dams?"; Woelfle-Erskine, "Connecting Rain to Taps and Drains to Gardens"; Woelfle-Erskine, "Collaborative Approaches to Flow Restoration in Intermittent Salmon-Bearing Streams"; Woelfle-Erskine, "Rain Tanks, Springs, and Broken Pipes as Emerging Water Commons along Salmon Creek, CA, USA"; Woelfle-Erskine, "Thinking with Salmon about Rain Tanks"; Deen et al., "Thirsty for Justice."
76 Puar, "I Would Rather Be a Cyborg than a Goddess."

Underflow 1

1 Sarna-Wojcicki, "Scales of Sovereignty," 314–15.
2 Browning, "Unearthing Subterranean Water Rights"; Nokes, "An Opportunity to Protect."
3 For example, Stanford and Ward, "An Ecosystem Perspective of Alluvial Rivers"; Boulton, "Hyporheic Rehabilitation in Rivers"; Power, Parker, and Dietrich, "Seasonal Reassembly of a River Food Web"; Datry, Larned, and Tockner, "Intermittent Rivers"; Nichols et al., "Water Temperature Patterns below Large Groundwater Springs."
4 Woelfle-Erskine, "Collaborative Approaches to Flow Restoration in Intermittent Salmon-Bearing Streams"; Woelfle-Erskine, Larsen, and Carlson, "Abiotic Habitat Thresholds for Salmonid Over-Summer Survival in Intermittent Streams"; Larsen and Woelfle-Erskine, "Groundwater Is Key to Salmonid Persistence and Recruitment in Intermittent Mediterranean-Climate Streams."

1. Thinking with Salmon about Water

1 Also key in this regulatory story is the process initiated by the North Coast Regional Water Quality Control Board in 2006 to regulate sediment and temperature in the Scott River. This process, known colloquially as the "TMDL Action Plan," is triggered by a listing as impaired under section 303(d) of the US Environmental Protection Agency's Clean Water Act.
2 In this account of the case, and in the Scott Valley case study later in the chapter, I draw on Dan Sarna-Wojcicki's analysis of the Scott Valley groundwater and salmon conflicts (Sarna-Wojcicki, "Scales of Sovereignty" and our coauthored account of multispecies and hyporheic imaginaries in the Scott Valley and Salmon Creek watersheds (Woelfle-Erskine and

Sarna-Wojcicki, in review). The directive refers to §1602 of the California Fish and Game Code.
3 Siskiyou Co. Farm Bureau v. Dept. Fish & Wildlife 06042015 . Specifically, the California Department of Fish and Wildlife could not require "notification of the act of extracting water pursuant to a valid water right where there is no alteration to the bed, bank, or stream."
4 California Department of Fish and Wildlife, *Lake and Streambed Alteration Program*.
5 Ostrom, *Governing the Commons*; Agrawal, "Sustainable Governance of Common-Pool Resources."
6 Todd, "Fish Pluralities," 105–6)
7 Agrawal, *Environmentality*; Peet and Watts, *Liberation Ecologies*.
8 Neimanis, Garrard, and Kerridge, *Bodies of Water*, 156. Neimanis, Garrard, and Kerridge write of different water imaginaries, "All imaginaries are a congeries of matter and meaning—ideas entangled with material situations that offer various orientations towards thickly emergent worlds."
9 Dawson, *Soldier Heroes*, 48; For another example, see Anna Tsing on how locally grounded concepts of "biodiversity" interact with, shape, and are reshaped by conversations in international NGO forums, in *Friction*.
10 Groves, *The Geological Unconscious*; Gómez-Barris and Joseph, "Coloniality and Islands"; Khan, "At Play with the Giants"; Brown, "Learning to Read the Great Chernobyl Acceleration"; Kirksey, Shapiro, and Brodine, "Hope in Blasted Landscapes"; Linton, *What Is Water?*; Neimanis, Garrard, and Kerridge, *Bodies of Water*.
11 My method and focus differ from those of Neimanis, Garrard, and Kerridge and Linton, with whom I share broad goals and politics. Neimanis, Garrard, and Kerridge use archival methods and focuses mostly on issues of drinking water supply, though she also engages anticolonial poetry, art, and political action for water and rivers in the chapter referenced above. Linton, too, foregrounds archival methods and the science, policy, and engineering concerned with building dams and diverting rivers for municipal supply. My methods are ethnographic and participatory-ecological and are focused on the politics and practice of riverine field sciences.
12 Jasanoff and Kim, in "Sociotechnical Imaginaries and National Energy Policies," offer a slightly different take on imaginaries of energy policy, which I find useful in its attention to power asymmetries rooted in colonialism, capitalism, and historical and ongoing resource extraction that shape communities' capacity to do science and make politics. Sociotechnical imaginaries also articulate apparatuses as collections of technologies, sciences, devices, and regulatory regimes but without a feminist orientation toward the body and affect.
13 Barad, *Meeting the Universe Halfway*. Barad uses the scanning tunneling microscope in a quantum physics lab and relations in a Calcutta jute

mill as examples of apparatuses and shows how different scientific questions in the lab or production imperatives in the jute mill give rise to different configurations of singularities such as people, equipment, government policies, and lands where jute is grown or uranium is mined.

14 Barad, "Intra-actions"; Barad, *Meeting the Universe Halfway*. Critical physical geographers and others have also explored these issues—who does science, with what budget and instrumentation, with what study site, funding, and political commitments—and I contribute this literature as well. See, for example, chapters 1 and 3 in Lave, Biermann, and Lane, "Introducing Critical Physical Geography."

15 See, for example, Sarna-Wojcicki, "Scales of Sovereignty." Sarna-Wojcicki describes how the Karuk Tribe commissioned a hydrological model after a UC Davis model developed in close collaboration with settler ranchers determined that the aquifer was not overdrafted. In another case of tribal–settler rancher conflict over salmon habitat that Breslow analyzes, Skagit Valley farmers commissioned salmon habitat studies after the Swinomish Tribe's data showed that tide gates and levees threatened salmon recovery there. Breslow, "Tribal Science and Farmers' Resistance."

16 Such imaginaries are a form of situated knowledge in that they are contextual and limited, rather than abstract, singular, and unbiased. As such, this knowledge yields partial perspectives that can together craft a stronger objectivity. Haraway, "Situated Knowledges"; Harding, "'Strong Objectivity.'"

17 Linton, *What Is Water?*, 14.

18 White, *The Organic Machine*, 111. In figuring the Columbia as an organic machine, White challenges human-nature binaries that have led twentieth-century river managers to treat the river as a purely mechanical and controllable entity, "a machine that can be disassembled and redesigned largely at will, as if its various parts can be assigned different functions with only a technical relation to other parts and functions."

19 White, *The Organic Machine*, 111–12.

20 Haraway, "A Manifesto for Cyborgs."

21 Bellacasa, *Matters of Care*.

22 Woelfle-Erskine, "Thinking with Salmon about Rain Tanks"; Coulthard, *Red Skin, White Masks*, 12.

23 Gardner and Clancy, "From Recognition to Decolonization."

24 For a feminist-phenomenological reading of such material transits through water-shaped bodies, see Neimanis, Garrard, and Kerridge, *Bodies of Water*.

25 The organizations include the Salmon Creek Watershed Council, the Klamath Basin Tribal Water Quality Work Group, the Scott River Watershed Council, and the Scott Valley Groundwater Advisory Council. These

agencies include the National Marine Fisheries Service, the California Department of Fish and Wildlife, Resource Conservation Districts, and the State and Regional Water Quality Control Boards.

26 Simpson, "Indigenous Resurgence and Co-resistance"; Coulthard and Simpson, "Grounded Normativity/Place-Based Solidarity."

27 Barad, "TransMaterialitiesTrans*/Matter/Realities and Queer Political Imaginings." Barad argues explicitly in works written after *Meeting the Universe Halfway* that imaginaries are material aspects of apparatuses. Like Barad, I think that imaginaries are material, or at least so tightly bound up in the actions people take to manage rivers and divert water that they effect action in the macroscopic material world.

28 Woelfle-Erskine, "Rain Tanks, Springs, and Broken Pipes as Emerging Water Commons along Salmon Creek, CA, USA."

29 For a detailed description of this process, see Woelfle-Erskine, "Collaborative Approaches to Flow Restoration in Intermittent Salmon-Bearing Streams."

30 Worster, "Hydraulic Society in California"; Worster, *Rivers of Empire*; Reisner, *Cadillac Desert*; Woelfle-Erskine et al., *Dam Nation*.

31 Linton and Budds, "The Hydrosocial Cycle"; Linton, *What Is Water?*; D'Souza, "Framing India's Hydraulic Crises"; King, *First Salmon*; Katz et al., "Impending Extinction of Salmon, Steelhead, and Trout (*Salmonidae*) in California"; Flagg, "Balancing Conservation and Harvest Objectives."

32 California Department of Water Resources, "California Water Plan Update 2005," 6.

33 California Water Code, Division 6, Part 2.74; Cantor et al., *Navigating Groundwater–Surface Water Interactions under the Sustainable Groundwater Management Act*, 5, referencing California Water Code § 10721.

34 Deen et al., "Thirsty for Justice"; Community Water Center, "Water Governance."

35 Author's notes, interview with Salmon Creek resident, June 12, 2012.

36 Over the seven years of my field work in Salmon Creek, which began at the beginning of the drought and continued for two years after it officially ended, I conducted thirty formal interviews and had informal conversations with an additional forty residents about their water supply. In the formal interviews, only two interviewees described having reliable water.

37 A year-class is a distinct subpopulation of an anadromous fish that hatches in a given year. Coho spawn three years after they hatch. In coho salmon and other anadromous fishes that have high temporal fidelity in their life history, a year-class that becomes locally extinct can only recover if spawning fishes stray into the depopulated stream or are reintroduced by humans.

38 Hammack, Prunuske, and Choo, "Salmon Creek Estuary Study"; Hammack, Hulette, and Prunuske, "Salmon Creek Integrated Coastal Watershed

Management Plan"; Prunuske Chatham Inc., Porter, and OAEC WATER Institute, *Salmon Creek Water Conservation Plan*.

39   This release is part of the Russian River Captive Broodstock Program, an experimental collaboration between various state and federal agencies in which individual salmon are raised to adulthood in a hatchery, then bred to maximize genetic diversity.

40   M. Fawcett, personal communication, August 8, 2012.

41   For example, the cost of improving the small Bodega Water Company or trucking in water.

42   Ostrom, *Governing the Commons*.

43   Dolman and Lundquist, *A Citizen's Guide to Protecting and Restoring Our Watersheds*.

44   Foglia et al., "Scott Valley Integrated Hydrologic Model."

45   Foglia et al., "Scott Valley Integrated Hydrologic Model."

46   Van Kirk and Naiman, "Relative Effects of Climate and Water Use on Base-Flow Trends in the Lower Klamath Basin."

47   The Quartz Valley Indian Reservation (QVIR) has treaty rights that it is leveraging to comanage surface and groundwater flows on Shackleford Creek, which flows through the reservation. QVIR also has fishing rights to salmon that spawn throughout the Scott Valley, as do the Karuk, Yurok, and Hoopa Tribes, who harvest fishes that may have reared in the Scott Valley. The Klamath Tribes, upstream of the four dams that currently block fish passage, also have a stake in Scott Valley flow and habitat conditions. A coho that reared in the Scott could, once the dams come down, travel further upriver to spawn in a different tributary, potentially in Klamath Tribes' fishing areas.

48   The Scott Valley contains the best coho salmon habitat in the Klamath Basin below the dams and is crucial to recovery.

49   In 1953, only 3 percent of irrigation water came from groundwater pumping. By 2001, 80 percent of irrigation water did.

50   Wells, *History of Siskiyou County, California*, 44.

51   Pollock, Beechie, and Jordan, "Geomorphic Changes Upstream of Beaver Dams in Bridge Creek, an Incised Stream Channel in the Interior Columbia River Basin, Eastern Oregon."

52   Woelfle-Erskine and Sarna-Wojcicki, "Dam 'Em All"; Woelfle-Erskine and Sarna-Wojcicki, "Hyporheic Imaginaries"; Sarna-Wojcicki and Woelfle-Erskine, "The Manifest Reversals of Multi-species Collaborative Watershed Restoration."

53   Author's notes, interviews conducted at the State of the Beaver conference, Canyonville, OR, February 2011, and in the field in the Scott Valley (CA), Yakima Valley (WA), and Methow Valley (WA).

54   From the following letters in the public comments on the groundwater study plan: Klamath Riverkeeper, Pacific Coast Federation of Fishermen's Associations, local resident.

55 I engage Glen Coulthard's critique of commons as erasing Indigenous sovereignties and survivance and ways that multispecies commons might incorporate these critiques in Woelfle-Erskine, "Beavers as Commoners."

56 Under the Clean Water Act, the Scott River and its tributaries get too warm in the summer for salmon; the allowed amount of a pollutant, including temperature, is known as the total maximum daily load, or TMDL. If a stream is found to exceed a TMDL standard, it is in violation of section 303(d) of the Clean Water Act; in California, the Regional Water Quality Control Board is responsible for studying the problem, proposing a solution, and monitoring temperatures to ensure the solution actually fixes the problem. According to Donald Coates et al., *Staff Report for the Action Plan for the Scott River Watershed Sediment and Temperature Total Maximum Daily Loads*, the primary factor affecting stream temperatures in the Scott River watershed is reductions of shade provided by near-stream vegetation. Other factors that impact water temperature include changes in groundwater accretion, where diversions of surface water have the potential to affect temperatures in smaller tributaries where the volume of water diverted is relatively large compared to the total stream flow; microclimate alterations resulting from near-stream vegetation removal; and changes in channel geometry from natural conditions.

57 Foglia et al., "Scott Valley Integrated Hydrologic Model"; Foglia, McNally, and Harter, "Coupling a Spatiotemporally Distributed Soil Water Budget with Stream-Depletion Functions to Inform Stakeholder-Driven Management of Groundwater-Dependent Ecosystems"; Foglia et al., "Modeling Guides Groundwater Management in a Basin with River-Aquifer Interactions"; Cantor et al., "Navigating Groundwater–Surface Water Interactions under the Sustainable Groundwater Management Act; Crystal Robbins, environmental director, Quartz Valley Indian Community, September 26, 2018.

58 Klamath Riverkeeper, "Comments to the Scott Valley Groundwater Study Plan."

59 Hydrologists call this stream state a "gaining stream."

60 Hydrologists call this stream state a "losing stream."

61 Quartz Valley and the Pacific Coast Federation of Fishermen's Associations also presented stream gauge data to support this model.

62 Harris, "2013–15 Three Year Summary."

63 Regarding Shackleford Creek, the Water Trust proposed that the stream had historically gone dry in late summer, an assertion that Quartz Valley dismissed, citing historical records and Indigenous knowledge.

64 Schrader shows how the ontological indeterminacy of a phenomenon (in this case the lifecycle of *Pfiesteria*) is resolved differently, depending on which agential cuts scientists make. She argues that this has profound

policy implications. Schrader, "Responding to *Pfiesteria piscicida* (the Fish Killer)."

65 USGS streamflow data and data from the state monitoring wells. Politics precluded full sharing of data—the Scott Valley ranchers agreed to share their data, confidentially, with the UC Davis researchers but not with Quartz Valley or other parties.

66 Klamath Riverkeeper, "Comments to the Scott Valley Groundwater Study Plan."

67 Foglia et al., "Modeling Guides Groundwater Management in a Basin with River-Aquifer Interactions"; Cantor et al., "Navigating Groundwater–Surface Water Interactions under the Sustainable Groundwater Management Act"; Scott Valley Groundwater Action Committee, *Voluntary Groundwater Management & Enhancement Plan for Scott Valley Advisory Committee (Approved 10-22-12)*.

68 See, for example, Table 2 in Cantor et al., "Navigating Groundwater–Surface Water Interactions under the Sustainable Groundwater Management Act."

69 Armstrong, "Comments to the Scott Valley Groundwater Study Plan." As of this writing (February 2019), data confidentiality was still built into monitoring proposals under the Sustainable Groundwater Management Act; see meeting agenda for the Groundwater Sustainability Agency, January 22, 2019, available here: www.co.siskiyou.ca.us/naturalresources/page/scott-valley-groundwater-advisory-committee-meeting.

70 For example, Simpson, "Settlement's Secret"; Tuck and Yang, "Decolonization Is Not a Metaphor"; Simpson, "Indigenous Resurgence and Co-resistance."

71 There was some convergence between these two positions around floodplain reconnection (where it doesn't conflict with infrastructure), work in the tailings reach, and beaver reintroduction, as Sarna-Wojcicki and I discuss in detail. Sarna-Wojcicki, "Scales of Sovereignty"; Woelfle-Erskine and Sarna-Wojcicki, "Hyporheic Imaginaries."

72 Foglia et al., "Scott Valley Integrated Hydrologic Model."

73 Water Education Foundation, "Overdraft."

74 Whyte, *Indigeneity and US Settler Colonialism*, 100. See also TallBear, "Genomic Articulations of Indigeneity."

75 Kormann, "How Carbon Trading Became a Way of Life for California's Yurok Tribe"; Karuk Tribe, "Land Management"; Johannson, "Warner Springs Ranch Comes Full Circle with Pala Purchase"; Associated Press, "California City Returns Island Taken from Native Tribe in 1860 Massacre"; Sogorea Te Land Trust, "Shuumi Land Tax"; Hannibal, "The Amah Mutsun and the Recovery of Traditional Ecological Knowledge"; Donna Martinez, "Land Returned to Pomo Tribe Forms Kashia Coastal Reserve"; Esselen Tribe of Monterey County, "Our History"; Koran, "Northern California Esselen Tribe Regains Ancestral Land after 250 Years."

76 Coulthard and Simpson write,

> Attacking the relationality of Indigenous political orders through the strategic targeting of Indigenous peoples' relationship to land has been a site of intense white supremacy and heteropatriarchy, serving as a mechanism to submit Indigenous lands and labor to the demands of capitalist accumulation and state-formation. Historically, Indigenous peoples have responded to this violence and negation through fierce and loving mobilization. Indigenous resistance and resurgence in response to the dispossessive forces of settler colonization, in both historical and current manifestations, employ measures and tactics designed to protect Indigenous territories and to reconnect Indigenous bodies to land through the practices and forms of knowledge that these practices continuously regenerate. What we are calling "grounded normativity" refers to the ethical frameworks provided by these Indigenous place-based practices and associated forms of knowledge. Grounded normativity houses and reproduces the practices and procedures, based on deep reciprocity, that are inherently informed by an intimate relationship to place. Grounded normativity teaches us how to live our lives in relation to other people and nonhuman life forms in a profoundly nonauthoritarian, nondominating, nonexploitive manner. Grounded normativity teaches us how to be in respectful diplomatic relationships with other Indigenous and non-Indigenous nations with whom we might share territorial responsibilities or common political or economic interests. Our relationship to the land itself generates the processes, practices, and knowledges that inform our political systems, and through which we practice solidarity.

Coulthard and Simpson, "Grounded Normativity/Place-Based Solidarity," 254.

77 Coulthard and Simpson, "Grounded Normativity/Place-Based Solidarity"; Kauanui, "'A Structure, Not an Event.'" The Federated Indians of Graton Rancheria is a federally recognized Native nation made up of Southern Pomo and Coast Miwok people with headquarters; the tribe was "terminated" (legally derecognized) by the US government in the 1950s and only regained federal recognition in 2002. Federated Indians of Graton Rancheria, "History."

78 Kauanui, "'A Structure, Not an Event.'"

79 Woelfle-Erskine, "Logjams, Check Dams, and the Gift of Foresight."

80 Bowman, "Beaver Valley."

81 Bowman, "Beaver Valley."

82 Bowman, "Beaver Valley."

83 Bowman, "Beaver Valley."

84 Scott River Watershed Council, *Restoring Priority Coho Habitat in the Scott River Watershed Modeling and Planning Report*.

85 psmfcvideo, *Beavers on Working Lands Featuring Landowners Betsy and Michael Stapleton.*

2. Queer × Trans × Feminist Ecology

1   Kafer, *Feminist, Queer, Crip.*
2   Samuels, "Six Ways of Looking at Crip Time."
3   Muñoz, *Cruising Utopia*, 182; Nyong'o, "Brown Punk."
4   Muñoz, "Feeling Brown, Feeling Down," 676.
5   Cruikshank, *Do Glaciers Listen?*; Tsing, "Arts of Inclusion, or How to Love a Mushroom."
6   Cid and Bowser, "Breaking Down the Barriers to Diversity in Ecology."
7   We Who Feel Differently was the title of a queer affect conference held at the New Museum in New York City on May 8, 2013, where Muñoz delivered a keynote on the brown commons and inspired my formulation of trans field affect here.
8   Anzaldúa, *Light in the Dark/Luz en lo oscuro*; Other feminist and trans writing in this vein includes Anzaldúa, *The Gloria Anzaldúa Reader*; Sedgwick, "A Poem Is Being Written"; Sedgwick, "White Glasses"; TallBear, *Making Love and Relations beyond Settler Sexualities*; Todd, "From a Fishy Place"; Todd, "Fish, Kin and Hope."
9   Muñoz, *Cruising Utopia.*
10 Sedgwick, *Tendencies.*
11 Butler, *Precarious Life*; Butler, *Frames of War.*
12 Muñoz, *Cruising Utopia*, 3.
13 Rose, "Double Death." In a piece that blends writing and imagery of dead and decomposing animals killed by humans or human artifacts, Bird Rose highlights interconnectedness among humans and other species and places as they ramify for life, culture, and time:

> So many losses occur that damaged ecosystems are unable to recuperate their diversity. The death of resilience and renewal, at least for a while. So many extinctions that the process of evolution is unable to keep up. The death of evolution itself, at least for a while. The unmaking of country, unraveling the work of generation upon generation of living beings; cascades of death that curtail the future and unmake the living presence of the past. The death of temporal, fleshy, metabolic relationships across generations and species. The destruction of the future of one's own death, which starts to collapse along with the future of flourishing others and ecosystems.

14 In *Ecologies of Comparison*, Tim Choy writes of Hong Kong residents who may in the future be displaced by development and their feelings and politics around displacement. He terms their feeling of loss for their home places "anticipatory nostalgia."

15 Muñoz, *Cruising Utopia*, 182.
16 Nyong'o, "Brown Punk," 84.
17 Muñoz, "After Jack."
18 Juno Parreñas's argument that orangutan sanctuaries provide species-level palliative care is one such proposition.
19 See, for example, the interviews with movement activists collected in Adamson, Evans, and Stein, *The Environmental Justice Reader*, as well as the community case studies in Deen et al., *Thirsty for Justice*.
20 For example, Nadasdy, *Hunters and Bureaucrats*; Sze et al., "Defining and Contesting Environmental Justice"; Wilkinson, *Messages from Frank's Landing*.
21 Muñoz, "Feeling Brown, Feeling Down."
22 In the digital sound atlas *What Is Missing?* Maya Lin solicited sounds of species that were near extinction or and earth elementals (like dammed rivers) that no longer existed. The title of this conclusion riffs on that project.
23 Muñoz, *Cruising Utopia*, 147.
24 Muñoz, *Cruising Utopia*, 147.
25 Hazard, *The Gold Fish, or, Straight Flushes for the Manifestly Destined*.
26 Harney and Moten, *The Undercommons*; for queer strategies for resisting assimilation, see also Sycamore, *That's Revolting!*
27 Whyte, "Food Sovereignty, Justice and Indigenous Peoples: An Essay on Settler Colonialism and Collective Continuance."

Underflow 3

1 Woelfle-Erskine, "Who Needs Dams?"
2 Cole, "Wash Out."
3 Hayward, "Lessons from a Starfish."
4 Hayward, "Lessons from a Starfish," 254.

3. The Watershed Body

1 For an accessible treatment of beaver restoration, including case studies from across the US, see Castro et al., *The Beaver Restoration Guidebook*.
2 Power et al., "Challenges in the Quest for Keystones."
3 Woelfle-Erskine and Cole, "Transfiguring the Anthropocene." The idea of "thinking with" an animal, plant, or other species is common in the field of critical animal studies and feminist science studies but perhaps unfamiliar to other readers. As I use it, the practice involves holding in your mind a concept—here, engineering—together with your sense of an organism (or perhaps an ecosystem, or elemental like fire or water). It helps, sometimes, to go to the field to observe that animal or its traces—this can help you avoid traps of symbolizing animals rather than perceiving

them and their relations more clearly. For other examples of this practice, consider Hustak and Myers on orchids, pollination, and plant sciences. Hustak and Myers, "Involutionary Momentum."

4   Martinez, "Protected Areas, Indigenous Peoples, and the Western Idea of Nature"; Norgaard et al., *Karuk Tribe Climate Vulnerability Assessment*; Marks-Block, Lake, and Curran, "Effects of Understory Fire Management Treatments on California Hazelnut, an Ecocultural Resource of the Karuk and Yurok Indians in the Pacific Northwest"; Whyte, "Our Ancestors' Dystopia Now"; Tulalip Tribes Natural Resources, "Beaver"; Mid-Columbia Fisheries Enhancement Group, "Yakima Beaver Project."
5   Woelfle-Erskine and Cole, "Transfiguring the Anthropocene."
6   Woelfle-Erskine and Cole, "Transfiguring the Anthropocene."
7   Woelfle-Erskine and Cole, "Transfiguring the Anthropocene," 308; Hayward, "Lessons from a Starfish," 256.
8   However, a few beavers have been spotted in recent years in Sonoma and Mill Creek, a few miles further inland.
9   National Marine Fisheries Service, *Final Recovery Plan for Central California Coast Coho Salmon Evolutionarily Significant Unit*; National Marine Fisheries Service, *Southern Oregon Northern California Coast Coho Salmon Recovery Plan*.
10 UC Cooperative Extension, "Russian River Coho Salmon Recovery Program."
11 These research questions emerged in conversations with biologists restoring Russian River habitat and with grassroots activists who had convinced scientists to designate Salmon Creek as critical habitat for coho under the Endangered Species Act.
12 David, Asarian, and Lake, "Wildfire Smoke Cools Summer River and Stream Water Temperatures."
13 Michael Fawcett, Lauren Hammack, and Brian Cluer, personal communication, November 20, 2012.
14 Larsen and Woelfle-Erskine, "Groundwater Is Key to Salmonid Persistence and Recruitment in Intermittent Mediterranean-Climate Streams."
15 Darwin, "On the Three Remarkable Sexual Forms of *Catasetum tridentatum*, an Orchid in the Possession of the Linnean Society."
16 Here I'm thinking with Hustak and Myers, "Involutionary Momentum," specifically their concept of "involutionary momentum," which describes the shape of the orchid and the gesture it makes to its pollinator in terms of desire—a desire that Darwin takes up in his description of how the bee pollinates the orchid and that he transmits to his reader. Hustak and Myers contrast this affective, desireful way of doing science with contemporary chemical ecologists who reduce plant communication to a mechanistic, self-interested process.
17 Largely due to Lundquist and collaborator Brock Dolman's work (including two papers published in CDFW's journal, *California Fish and Game*,

and numerous white papers and meetings), the agency's policy is shifting.
18 Kirksey, Shapiro, and Brodine, "Hope in Blasted Landscapes."
19 Muñoz, "Gestura, Ephemera, and Queer Feeling," 81.
20 First published as "Gesture, Ephemera, and Queer Feeling: Approaching Kevin Aviance," in *Dancing Desires*, ed. Jane Desmond (Hanover: Wesleyan University Press, 2001), this chapter also appears in Muñoz, *Cruising Utopia*.
21 Muñoz, "Gestura, Ephemera, and Queer Feeling: Approaching Kevin Aviance," 73–74.
22 Muñoz, "Ephemera as Evidence," 6.
23 The first staged reading took place at Bluestockings Books in New York City in 2007. The full musical version was staged in 2011 and 2012; Sarolta Jane Vay directed a 2018 film adaptation, *The Gold Fish Casino*.
24 Dellecave, *Nocturnal Beaver*. From the artist's website: "Nocturnal Beaver was a 5-day endurance installation and accumulating sculpture that combined notions of ecology, feminism, body art, endurance performance, collective work, disruption, and radical sex-positivity."
25 Dellecave, *Nocturnal Beaver*.
26 Dellecave, *Nocturnal Beaver*.
27 Ezra Berkley Nepon, *No One Mourns The Wicked/The Wizard of Shushan*, Purim 2017, Brooklyn, NY, March 17, 2017, https://ezraberkleynepon.com/2017/03/17/purim2017.
28 Woelfle-Erskine and Cole, "Transfiguring the Anthropocene."
29 Muñoz, "Gestura, Ephemera, and Queer Feeling," 81.
30 Muñoz, "Gestura, Ephemera, and Queer Feeling," 81.
31 See, for example, Subramaniam, *Ghost Stories for Darwin*. Subramaniam theorizes from the case of a chaparral restoration project outside Los Angeles, in which scientists rejected the concepts of "alien" and "invasive" species, instead describing in nuanced language how different plants influenced each other given different fire and rainfall patterns. These more nuanced descriptions of plants' ecological dynamics pointed to new options for human intervention.
32 Muñoz, *Cruising Utopia*, 74.
33 Castoreum is the clear, strongly scented fluid that beavers use to mark territory and recognize one another. They expel it from a gland located inside their anus; beaver live-trappers express some of it to identify beavers by sex.
34 My analysis here is based on content analysis of thirty-five interviews conducted between 2011 and 2014 at practitioner conferences (Salmonid Restoration Federation and State of the Beaver) and field visits and by phone. I conducted some of this fieldwork collaboratively with Dan Sarna-Wojcicki.

35 Worth a Dam, Worth a Dam—The Martinez Beavers; Pollock et al., "Using Beaver Dams to Restore Incised Stream Ecosystems."
36 Tsing, "Arts of Inclusion, or How to Love a Mushroom."
37 Tsing, *The Mushroom at the End of the World*.
38 Muñoz, "Feeling Brown, Feeling Down."
39 As USFS biologist Kent Woodruff told me, some even stop calling his crew to live trap, because they have come to appreciate beavers' river modifications.
40 Koenigsberg, *The Beaver Believers*.
41 Author's interview notes, interview with Michael Pollock, State of the Beaver Conference, February 11, 2011.
42 Woelfle-Erskine and Cole, quoting Hayward, "Lessons from a Starfish," and Haraway, "Foreword," xxiv.
43 Muñoz, "Gestura, Ephemera, and Queer Feeling," 81.
44 Muñoz, "Gestura, Ephemera, and Queer Feeling," 79.
45 Muñoz, "Gestura, Ephemera, and Queer Feeling," 79.
46 Later, I used field ecology as a grounds for queer, trans, and feminist theory (see chapter 2) and more recently I am using trans and queer theory to develop new practices of teaching and research in field ecology.
47 Woelfle-Erskine and Cole, "Transfiguring the Anthropocene, 307.
48 This story is foundational to the self-appointed "beaver believer" movement—a loose network of biologists, geomorphologists, and watershed advocates who want to reintroduce beavers as a way to increase streamflow, create juvenile salmon habitat, and recharge groundwater. As such, the story has circulated through salmon-restoration networks across the Pacific West and in beaver advocacy networks that span the Atlantic. I first heard one version of it from Brock Dolman, a beaver advocate who visited this dam later with Will Harling and had his own moment of reckoning with settler legacies of beaver removal while overlooking this beaver pond. I heard another version from my collaborator Dan Sarna on a long drive back from the Klamath River, where our plans to install posts to support a beaver dam on Seiad Creek were thwarted by a forest fire.
49 Harling, "One Beaver at a Time."
50 Chen, *Animacies*. Such inversions of animacy hierarchies—ascribing humanlike agency to rocks, landforms, and animals—are what Mel Chen describes as a queer move of undoing normal categories and boundaries in order to upend racial and colonial hierarchies.
51 Harling, "One Beaver at a Time."
52 Harney and Moten, *The Undercommons*, 73.
53 Harney and Moten, *The Undercommons*, 74.
54 Harling, "One Beaver at a Time."
55 Hayward, "Lessons from a Starfish," 255.
56 For example, cultural fire and elk studies funded by the Forest Service and carried out by the Yurok and Karuk natural resources departments,

off-channel salmon ponds funded by a hydroelectric company and the US Fish and Wildlife Service and constructed by the Mid-Klamath Watershed Council and the Karuk fisheries department, and food sovereignty work funded by the USDA and carried out by UC Berkeley and Karuk and Yurok natural resources departments.
57 Tsing, *The Mushroom at the End of the World*.
58 Bellacasa, "'Nothing Comes without Its World,'"212.
59 Kimmerer, "The Covenant of Reciprocity"; Whyte, "Food Sovereignty, Justice and Indigenous Peoples."
60 This future-dream was inspired by Cole, "Wash Out," specifically prophecies Clovis and Larch unearthed in Hazard's Unmoored Frontier.

Underflow 4

1 Wilson, *Research Is Ceremony*, 7.
2 Tafoya, "Finding Harmony."
3 Wilson, *Research Is Ceremony*, 7.
4 My understanding of family and kinship here is strongly influenced by conversations with Kim TallBear during reading for my qualifying exams, around thinking across queer and Indigenous notions of family, nature, and ecology. TallBear's thoughts on these topics are collected on her blog, TallBear, *The Critical Polyamorist*.

4. Unchartable Grief

1 Freeman, "Queer Belongings."
2 Woelfle-Erskine, Larsen, and Carlson, "Abiotic Habitat Thresholds for Salmonid Over-Summer Survival in Intermittent Streams," 3. To increase readability, I have removed citations that appear in the paper.
3 Woelfle-Erskine, Larsen, and Carlson, "Abiotic Habitat Thresholds for Salmonid Over-Summer Survival in Intermittent Streams," 19.
4 This question—on feeling too much or not enough—was posed by Katie Gillespie and Patricia Lopez in a call for papers for an anthology of writing on the politics of grief in the field. An earlier draft of this section was published in that volume, Gillespie and Lopez, *Vulnerable Witness*.
5 Here, I draw on my mostly US-based experience and on what I've gathered from colleagues and memoirs.
6 Murphy, "Unsettling Care," 721.
7 Ahmed, *The Promise of Happiness*.
8 Murphy, "Unsettling Care," 721.
9 Freeman, "Queer Belongings," 297.
10 Freeman, "Queer Belongings," 297.
11 Freeman, "Queer Belongings," 297–98.
12 Freeman, "Queer Belongings."

13 Freeman, "Queer Kinship," 299.
14 Tsing, "Arts of Inclusion, or How to Love a Mushroom."
15 Haraway, *When Species Meet*; Tsing, "Arts of Inclusion, or How to Love a Mushroom."
16 Freeman's gloss of Bourdieu's term introduces aspects of temporality that are important for my argument about more-than-human kinship, both in the example later in this chapter and in the beaver case developed in chapter 3. She writes, "Here, of course, 'generation' is a cultural and not a biological matter: the phrase evokes J. L. Austin's and Jacques Derrida's understanding of speech and even being itself as performances enacted under particular cultural constraints, but open to the possibility of improvisatory 'resignification.' But also, the phrase captures the dynamic between the stasis of waiting, which appears in the term 'durably,' and the mobility of 'improvisation' that is key to habitus. *Habitus* sediments practices that are, Bourdieu insists, temporally structured: practices are 'defined by the fact that their temporal structure, direction, and rhythm [are] *constitutive* of their meaning.'" Freeman, "Queer Belongings," 308–9.
17 Freeman, "Queer Belongings," 305–6.
18 Freeman, "Queer Belongings," 308.
19 Fabre, *Fabre's Book of Insects*.
20 Hustak and Myers, "Involuntary Momentum," 92.
21 Hustak and Myers, "Involuntary Momentum," 79.
22 Hustak and Myers, "Involuntary Momentum," 96–97.
23 Hustak and Myers, "Involuntary Momentum," 106.
24 Freeman, "Queer Belongings," 305
25 Rose, "Double Death."
26 Murphy, "Unsettling Care," 721.
27 Schuller, *The Biopolitics of Feeling*, 209.
28 Schuller, *The Biopolitics of Feeling*, 209.
29 Ferguson, "To Catch a Light-Filled Vision."

Underflow 5

1 See Ecopoetics along Shorelines, http://shorelinepoetics.com. The text in the remainder of this section is adapted from materials that Hazard and I developed for these classes and workshops.
2 Brainard, *I Remember*. For Blanchfield's version, which July Hazard encountered in Blanchfield's class at the University of Montana, column 1 contained pieces of the deliberately remembered moment, and column 2 contained the spontaneous, unrelated flashes of memories that arose alongside.
3 Barad, *Meeting the Universe Halfway*, 300.

5. With and for the Multitude

Unless otherwise specified, all quotes in this chapter are from my transcript of Muñoz, "The Brown Commons," and thus the citations lack page numbers.

1. Waterlines, "The Waterlines Project Map."
2. Muñoz, *The Sense of Brown*. At the time of Muñoz's passing in 2013 and my first drafts of this chapter in 2016, only two of his works on brown commons had been published. As I submitted the final manuscript, *The Sense of Brown* came out, but some of the text that I draw on here is available only as recordings of public talks or the Q&A.
3. See, for example, Delany, *Times Square Red, Times Square Blue*; Wojnarowicz, *The Waterfront Journals*; Rechy, *City of Night*; Arenas, *Before Night Falls*.
4. For an important discussion of the urban ecological potential of Delany's theoretical work on cruising, see Ensor, "Queer Fallout." Ensor writes that Muñoz drew from Delany theorizing cruising as a method and mode of thought.
5. Muñoz wrote about such queer genealogies in "'Gimme Gimme This . . . Gimme Gimme That'" (96) in relation to the poet Jack Spicer, who used the term to describe his own poetic lineage:

   > There are of course many queer genealogies, and they lead to sites and movements that are not always self-declared as queer or even LGBT. To think along these grooves we should resist the impulse to simply "queer" an object, phenomenon, or historical moment and instead attend to it with an understanding of lines of queer genealogical connectivity as something other than tautological. The challenge here is to look to queerness as a mode of "being-with" that defies social conventions and conformism and is innately heretical yet still desirous for the world, actively attempting to enact a commons that is not a pulverizing, hierarchical one bequeathed through logics and practices of exploitation.

6. See a partial reading list at Float the Blue Squealer, "Queer Theory Reading List," http://bluesquealer.wordpress.com; Muñoz, "The Brown Commons."
7. Harney and Moten, *The Undercommons*.
8. Harney and Moten, *The Undercommons*.
9. Muñoz, "The Brown Commons"; Muñoz et al., "Theorizing Queer Inhumanisms."
10. Muñoz, "Keynote"; Muñoz, "Feeling Brown"; Stanley, "The Affective Commons"; Muñoz, *Cruising Utopia*, preface to tenth-anniversary edition.
11. Byrd et al., "Predatory Value."
12. Muñoz, "The Brown Commons."
13. Muñoz, "The Brown Commons."

14 Muñoz, "The Brown Commons." Throughout this chapter, section titles (apart from "Queer Ecology and Environmental Justice") are taken from Muñoz, from the various sources I mention above.
15 Cole, "Sunshine and Other Queer Waste"; Dillon, "Race, Waste, and Space."
16 Muñoz, "The Brown Commons."
17 Ferguson, "To Catch a Light-Filled Vision"; Gordon, "Some Thoughts on Haunting and Futurity." Elsewhere, Gordon explains the term as an abolitionist tactic, of making new worlds within or despite the constraints of the present: "Abolition involves critique, refusal and rejection of that which you want to abolish, but it also involves being or 'becoming unavailable for servitude', to use Toni Cade Bambara's words. Needless to say, being or becoming unavailable for servitude takes a certain amount of time and trouble and one reason why is that, among other things, being or becoming unavailable for servitude involves cultivating an indifference, an ability to be in-difference to the system's own benefits and its own technologies of improvement."
18 Ferguson, "To Catch a Light-Filled Vision."
19 Coulthard and Simpson, "Grounded Normativity/Place-Based Solidarity."
20 Todd, "Fish, Kin and Hope."
21 Barad, "Intra-actions."
22 Muñoz, "Keynote." In the Q&A following this talk at the We Who Feel Differently symposium, Muñoz described queer ecology as "an urban ecology that contains the natural and the unnatural." He then paused, laughed a little, and finished, "The organic and inorganic." I'm interested in the slippage and movement from these terms, the ways in which they are not equivalent. *Unnatural* certainly signals queerness, and also perhaps the anthropogenic—buildings, bridges, power grids, and the like. Are urban ecologies always queer because there are always queers in them, or because human interventions into the organic and inorganic relations in the space have been so drastically altered by infrastructure and novel species introductions? Such questions merit further thought."
23 Ecologists will likely think here of disturbance theory, emergent properties of ecosystems, dynamic equilibria, and C. S. Holling's schema of resilience.
24 Spicer, "Improvisations on a Sentence by Poe."
25 Muñoz, "'Gimme Gimme This . . . Gimme Gimme That,'" 97–98.
26 Muñoz, "The Brown Commons"; Tsang, *Wildness*.
27 Muñoz, "'Gimme Gimme This . . . Gimme Gimme That,'" 109.
28 Tuck, "Breaking up with Deleuze," 416, cited in Tuck and Yang, "R-words," 243.
29 Whyte, "Food Sovereignty, Justice and Indigenous Peoples."
30 Duwamish River Cleanup Coalition, "Duwamish River Cleanup Coalition/Technical Advisory Group (DRCC/TAG)."
31 Tuck and Yang, "R-words," 236.

32 Chen, *Animacies*. I'm thinking here of mercury from the New Almaden mine flowing down concrete-lined creeks through the new boomtowns of Silicon Valley and into the South San Francisco Bay.
33 Dillon, "Race, Waste, and Space."
34 Ferguson, "To Catch a Light-Filled Vision."
35 Muñoz, "The Wildness of the Punk Rock Commons."
36 In the texts I'm working from, Muñoz mention animals only once and is not concerned with relations among species and earth processes, which ecology takes as its object of study.
37 Muñoz, *Cruising Utopia*. See also "Before and After," preface to the tenth-anniversary edition.
38 Halberstam and Nyong'o, "Introduction," 462.
39 Muñoz, "The Brown Commons."
40 Muñoz, "The Brown Commons."
41 "Muñoz, "Keynote." In the question and answer to his keynote, Muñoz clarified, "What I'm calling white affect is a shorthand. I don't want to map it onto an identity . . . it's a normative way of being in the world transmitted through mediatization, pop culture."
42 Muñoz, "Keynote." In response to an audience question about space, Muñoz said,

> Space is really important, I'm interested in real space, actual space, and symbolic spaces, and affective landscapes, the space of shared emotion and shared modes of feeling that cross space. Anna Tsing's concept of friction [is] another way of thinking about space. Friction that happens across different spaces, productive for different things. *Wildness* is an antirealist documentary. Instead of being a documentary about a gay bar, it does it differently by giving the bar a voice, talking about the ways that life happens around the bar. There are these beautiful heartbreaking sequences about one member of the community being killed in what seems like a hate crime, another being deported, so all these histories of spatial displacement all spiral out of the actual space. . . . There are so many slow shots that aren't just establishing shots, not about establishing the location of the bar, but really about the city itself, the bar's place in what I'm calling this queer ecology, which is an urban ecology which includes the natural and the unnatural, the organic and the inorganic.

43 Muñoz, "The Brown Commons."
44 Muñoz, "The Brown Commons."
45 Muñoz, "Feeling Brown."
46 Wiegman, "The Times We're In," 4.
47 Coulthard and Simpson, "Grounded Normativity/Place-Based Solidarity"; Coulthard, *Red Skin, White Masks*.
48 Muñoz, "Keynote."

49 Gardner and Clancy, "From Recognition to Decolonization." In Coulthard's *Red Skin, White Masks*, he briefly mentions this critique as well.
50 TallBear, "Uppsala 3rd Supradisciplinary Feminist Technoscience Symposium."
51 Duwamish River Cleanup Coalition, "Duwamish River Cleanup Coalition/Technical Advisory Group (DRCC/TAG)."
52 Muñoz, "The Brown Commons."
53 This quote is from the question-and-answer period following Muñoz, "Keynote." In response to a question about politics of difference, Muñoz said, "We want to think difference as something that is not who we are, but a byproduct of our situation. I'm always arguing for the moving from isolated individualized particularities to a better sense of the multiple. Of being a collective integer as opposed to being just an isolated one. It's part of my critical utopian thing—the idea isn't just to be content in our difference, but always to be becoming."
54 Muñoz, "The Brown Commons."
55 On the Duwamish, these include diesel particulates from trucking, ozone, gravel dust, factory emissions and seeps of chemicals, storm runoff, and toxic ecologies of mercury methylation and PCB mobility through the food web.
56 The Environmental Coalition of South Seattle, where two Doris Duke conservation scholars that I mentored worked as interns during the summer of 2018 and helped organize the water festival.
57 For example, Dillon, "War's Remains"; Kim, "Bring on the Yuppies and the Guppies!"; Krupar, *Hot Spotter's Report*.
58 Thrush, *Native Seattle*; Thrush, "City of the Changers."
59 Nyong'o, "Brown Punk."
60 Agard-Jones, "What the Sands Remember."
61 Muñoz, "The Brown Commons."
62 Lindsey Dillon helped me think through the question of whether the brown commons excludes white people or just whiteness, reminding me of the line in Muñoz's "Feeling Brown, Feeling Down," "The world of *The Sweetest Hangover* is a world without white people," 676. Muñoz makes it clear in Q&A sessions that the brown commons negates structures of white supremacy and affects of whiteness but could in some cases *include* non-Brown people.
63 Muñoz, "The Brown Commons."
64 Muñoz, "'Gimme Gimme This . . . Gimme Gimme That.'"
65 Muñoz, "The Brown Commons."

Epilogue

1 Whyte, *Indigeneity and US Settler Colonialism*, 1:96.
2 Byrd et al., "Predatory Value," 12.

3   Simpson, *Mohawk Interruptus*, cited in Coulthard and Simpson, "Grounded Normativity/Place-Based Solidarity," 254.
4   The podcast interview is with trans Karuk collaborator Jasmine Harvey; the reading group is organized by Shawn Bourque. Both are settlers with close ties to the Karuk Tribe who work as employees of the Karuk Department of Natural Resources.
5   In describing the blue-line river as a settler desire, I think with Tuck and Yang, "R-Words," on the politics of desire in research in Black and Indigenous communities; I respond more fully to their provocation elsewhere in this book.
6   For example, prioritizing floodplain land acquisition is the official policy of the Confederated Salish-Kootenai Tribes, the Stillaguamish Tribe, the Nisqually Tribe, and other Pacific Northwest tribes that work for salmon recovery.
7   Sarna-Wojcicki, Sowerwine, and Hillman, "Decentring Watersheds and Decolonising Watershed Governance."
8   The dams have changed hands several times, from PacifiCorp to Scottish Power, then back to PacifiCorp and Berkshire-Hathaway.
9   Woelfle-Erskine and Sarna-Wojcicki, "Hyporheic Imaginaries."
10 Woelfle-Erskine and Sarna-Wojcicki, "Dam 'Em All."
11 Woelfle-Erskine and Cole, "Transfiguring the Anthropocene."
12 Haraway, "Situated Knowledges"; Harding, "'Strong Objectivity'"; Harding, *Sciences from Below*; Muñoz, "The Brown Commons"; Hayward, "Lessons from a Starfish"; Stryker, "(De)Subjugated Knowledges"; Woelfle-Erskine and Cole, "Transfiguring the Anthropocene."
13 Subramaniam and Willey, "Introduction."

Bibliography

500 Queer Scientists. "500 Queer Scientists Visibility Campaign." 500 Queer Scientists. Accessed October 20, 2020. https://500queerscientists.com.
Adams, John Joseph, ed. *Loosed upon the World: The Saga Anthology of Climate Fiction*. New York: Saga, 2015.
Adamson, Joni, Mei Mei Evans, and Rachel Stein. *The Environmental Justice Reader: Politics, Poetics, and Pedagogy*. Tucson: University of Arizona Press, 2002.
Agard-Jones, Vanessa. "What the Sands Remember." *GLQ: A Journal of Lesbian and Gay Studies* 18, no. 2–3 (June 1, 2012): 325–46. https://doi.org/10.1215/10642684-1472917.
Agrawal, A. *Environmentality: Technologies of Government and the Making of Subjects*. Durham, NC: Duke University Press, 2005.
———. "Sustainable Governance of Common-Pool Resources: Context, Methods, and Politics." *Annual Review of Anthropology* 32 (2003): 243–62.
Ahmed, Sara. *The Promise of Happiness*. Durham, NC: Duke University Press, 2010.
Anzaldúa, Gloria. *The Gloria Anzaldúa Reader*. Durham, NC: Duke University Press, 2009.
———. *Light in the Dark/Luz en lo oscuro: Rewriting Identity, Spirituality, Reality*. Edited by AnaLouise Keating. Durham, NC: Duke University Press, 2015.
Arenas, Reinaldo. *Before Night Falls: A Memoir*. Translated by Dolores M. Koch. Reprint, London: Penguin Books, 1994.
Armstrong, Marcia. "Comments to the Scott Valley Groundwater Study Plan." October 2007. http://groundwater.ucdavis.edu/files/136428.pdf.
Associated Press. "California City Returns Island Taken from Native Tribe in 1860 Massacre." *Guardian*. October 21, 2019. www.theguardian.com/us-news/2019/oct/21/california-city-returns-island-taken-from-native-tribe-in-1860-massacre.
Balazs, Carolina L., and Rachel Morello-Frosch. "The Three Rs: How Community-Based Participatory Research Strengthens the Rigor, Relevance, and Reach

of Science." *Environmental Justice* 6, no. 1 (February 1, 2013): 9–16. https://doi.org/10.1089/env.2012.0017.

Barad, Karen. "Intra-actions." Interview by Adam Kleinman. *Mousse* 34 (2012).

———. *Meeting the Universe Halfway: Quantum Physics and the Entanglement of Matter and Meaning*. Durham, NC: Duke University Press, 2007.

———. "TransMaterialitiesTrans*/Matter/Realities and Queer Political Imaginings." *GLQ: A Journal of Lesbian and Gay Studies* 21, no. 2–3 (June 1, 2015): 387–422. https://doi.org/10.1215/10642684-2843239.

Beckett, Caitlynn, and Arn Keeling. "Rethinking Remediation: Mine Reclamation, Environmental Justice, and Relations of Care." *Local Environment* 24, no. 3 (March 4, 2019): 216–30. https://doi.org/10.1080/13549839.2018.1557127.

Bellacasa, María Puig de la. *Matters of Care*. 3rd ed. Minneapolis: University of Minnesota Press, 2017.

———. "'Nothing Comes without Its World': Thinking with Care." *Sociological Review* 60, no. 2 (May 1, 2012): 197–216. https://doi.org/10.1111/j.1467-954X.2012.02070.x.

Benjamin, Walter. "Theses on the Philosophy of History." In *Illuminations*, edited by Hannah Arendt, translated by Harry Zorn, 253–64. New York: Schocken Books, 1950.

Bersani, Leo. "Is the Rectum a Grave?" *October* 43 (1987): 197–222. https://doi.org/10.2307/3397574.

Boulton, Andrew J. "Hyporheic Rehabilitation in Rivers: Restoring Vertical Connectivity." *Freshwater Biology* 52, no. 4 (April 1, 2007): 632–50. https://doi.org/10.1111/j.1365-2427.2006.01710.x, 10.1111/j.1365-2427.2006.01710.x.

Bowman, John. "Beaver Valley." *Siskiyou Daily News*. May 10, 2013. www.siskiyoudaily.com/article/20130510/NEWS/130519963.

Brainard, Joe. *Joe Brainard: I Remember*. Edited by Ron Padgett. New York: Granary Books, 2001.

Breslow, Sara Jo. "Tribal Science and Farmers' Resistance: A Political Ecology of Salmon Habitat Restoration in the American Northwest." *Anthropological Quarterly* 87, no. 3 (September 7, 2014): 727–58. https://doi.org/10.1353/anq.2014.0045.

Brown, Kate. "Learning to Read the Great Chernobyl Acceleration: Literacy in the More-than-Human Landscapes." *Current Anthropology* 60, no. S20 (2019): S198–208. https://doi.org/10.1086/702901.

Browning, Jordan. "Unearthing Subterranean Water Rights: The Environmental Law Foundation's Efforts to Extend California's Public Trust Doctrine." *Environs: Environmental Law and Policy Journal* 34 (2010): 231.

Butler, Judith. *Frames of War: When Is Life Grievable?* London: Verso, 2016.

———. *Precarious Life: The Powers of Mourning and Violence*. Reprint, London: Verso, 2006.

Butler, Octavia E. *Parable of the Sower.* Updated, New York: Grand Central, 2000.
Byrd, Jodi A. *The Transit of Empire: Indigenous Critiques of Colonialism.* 4th ed. Minneapolis: University of Minnesota Press, 2011.
Byrd, Jodi A., Alyosha Goldstein, Jodi Melamed, and Chandan Reddy. "Predatory Value: Economies of Dispossession and Disturbed Relationalities." *Social Text* 36, no. 2 (135) (June 1, 2018): 1–18. https://doi.org/10.1215/01642472-4362325.
Cairns, John, and John R. Heckman. "Restoration Ecology: The State of an Emerging Field." *Annual Review of Energy and the Environment* 21, no. 1 (1996): 167–89. https://doi.org/10.1146/annurev.energy.21.1.167.
California Department of Fish and Wildlife. *Lake and Streambed Alteration Program.* California Department of Fish and Wildlife. Accessed March 8, 2019. www.wildlife.ca.gov/Conservation/LSA.
California Department of Water Resources. "California Water Plan Update 2005." Sacramento, CA: California Department of Water Resources, 2005. www.waterplan.water.ca.gov/previous/cwpu2005/index.cfm.
California State Water Resources Control Board. "Water Rights: Public Trust Resources." California State Water Resources Control Board. Accessed November 27, 2020. www.waterboards.ca.gov/waterrights/water_issues/programs/public_trust_resources/#beneficial.
Canavan, Gerry, and Kim Stanley Robinson, eds. *Green Planets: Ecology and Science Fiction.* Annotated, Middletown, CT: Wesleyan, 2014.
Cantor, Alida, Dave Owen, Thomas Harter, Nell Nylen, and Michael Kiparsky. *Navigating Groundwater–Surface Water Interactions under the Sustainable Groundwater Management Act.* Berkeley, CA: Berkeley Law, March 2018. www.law.berkeley.edu/research/clee/research/wheeler/gw-sw.
Carrera, Jennifer S., Kent Key, Sarah Bailey, Joseph A. Hamm, Courtney A. Cuthbertson, E. Yvonne Lewis, Susan J. Woolford, et al. "Community Science as a Pathway for Resilience in Response to a Public Health Crisis in Flint, Michigan." *Social Sciences* 8, no. 3 (March 2019): 94. https://doi.org/10.3390/socsci8030094.
Cashman, Suzanne B., Sarah Adeky, Alex J. Allen, Jason Corburn, Barbara A. Israel, Jaime Montaño, Alvin Rafelito, et al. "The Power and the Promise: Working with Communities to Analyze Data, Interpret Findings, and Get to Outcomes." *American Journal of Public Health* 98, no. 8 (August 2008): 1407–17. https://doi.org/10.2105/AJPH.2007.113571.
Castro, Janine, Michael Pollock, Chris Jordan, Gregory Lewallen, and Kent Woodruff. *The Beaver Restoration Guidebook: Working with Beaver to Restore Streams, Wetlands, and Floodplains.* Portland, OR: US Fish and Wildlife Service, 2017. www.fws.gov/oregonfwo/ToolsForLandowners/RiverScience/Documents/BRG%20v.1.0%20final%20reduced.pdf.
Celermajer, Danielle, David Schlosberg, Lauren Rickards, Makere Stewart-Harawira, Mathias Thaler, Petra Tschakert, Blanche Verlie, and Christine

Winter. "Multispecies Justice: Theories, Challenges, and a Research Agenda for Environmental Politics." *Environmental Politics* 30, nos. 1–2 (October 7, 2020): 1–22. https://doi.org/10.1080/09644016.2020.1827608.

Chen, Mel Y. *Animacies: Biopolitics, Racial Mattering, and Queer Affect*. Durham, NC: Duke University Press, 2012.

Choy, Timothy. *Ecologies of Comparison: An Ethnography of Endangerment in Hong Kong*. Durham, NC: Duke University Press, 2011.

Cid, Carmen R., and Gillian Bowser. "Breaking Down the Barriers to Diversity in Ecology." *Frontiers in Ecology and the Environment* 13, no. 4 (May 1, 2015): 179. https://doi.org/10.1890/1540-9295-13.4.179.

Coates, Donald, Elmer Dudik, Richard Fadness, Rebecca Fitzgerald, Ranjit Gill, Bruce Gwynne, David Leland, Bryan McFadin, Carey Wilder, and Ben Zabinsky. *Staff Report for the Action Plan for the Scott River Watershed Sediment and Temperature Total Maximum Daily Loads*. Santa Rosa: State of California North Coast Regional Water Quality Control Board, December 7, 2005. www.waterboards.ca.gov/northcoast/water_issues/programs/tmdls/scott_river/staff_report.

Cole [a.k.a. Hazard], July Oskar. *The Gold Fish, or, Straight Flushes for the Manifestly Destined*, Crocker Museum of Art, Sacramento, CA, October 16, 2010.

———. "Sunshine and Other Queer Waste." Undisciplined Environments conference, Stockholm, March 22, 2016.

———. "Wash Out: Fluvial Forms and Processes on the American Frontiers." Master's thesis, University of Montana, 2010.

Community Water Center. "Water Governance." Community Water Center. Accessed December 1, 2015. www.communitywatercenter.org/water_governance.

Corburn, Jason. "Community Knowledge in Environmental Health Science: Co-producing Policy Expertise." *Environmental Science & Policy* 10, no. 2 (April 2007): 150–61. https://doi.org/10.1016/j.envsci.2006.09.004.

———. *Street Science: Community Knowledge and Environmental Health Justice*. Cambridge, MA: MIT Press, 2005.

Coulthard, Glen. *Red Skin, White Masks*. Minneapolis: University of Minnesota Press, 2014. www.upress.umn.edu/book-division/books/red-skin-white-masks.

Coulthard, Glen, and Leanne Betasamosake Simpson. "Grounded Normativity/Place-Based Solidarity." *American Quarterly* 68, no. 2 (June 28, 2016): 249–55. https://doi.org/10.1353/aq.2016.0038.

Crimp, Douglas. "Mourning and Militancy." *October* 51 (1989): 3–18. https://doi.org/10.2307/778889.

Cronin, Amanda, and David M. Ostergren. 2007. "Tribal Watershed Management: Culture, Science, Capacity, and Collaboration." *American Indian Quarterly* 31 (1): 87–109.

Cruikshank, Julie. *Do Glaciers Listen? Local Knowledge, Colonial Encounters, and Social Imagination*. Vancouver: UBC Press, 2010.

Daigle, Michelle. "Resurging through Kishiichiwan." *Decolonization: Indigeneity, Education & Society* 7, no. 1 (2018): 159–72.

Daigle, Michelle, and Margaret Marietta Ramírez. "Decolonial Geographies." In *Keywords in Radical Geography: Antipode at 50*, edited by Antipode Editorial Collective, 78–84. Hoboken, NJ: John Wiley & Sons, 2019. https://doi.org/10.1002/9781119558071.ch14.

Darwin, Charles. "On the Three Remarkable Sexual Forms of *Catasetum tridentatum*, an Orchid in the Possession of the Linnean Society." *Proceedings of the Linnean Society of London Botany*, no. 6 (1862): 151–57.

Datry, Thibault, Scott T. Larned, and Klement Tockner. "Intermittent Rivers: A Challenge for Freshwater Ecology." *BioScience* 64, no. 3 (2014): 229–35. https://doi.org/10.1093/biosci/bit027.

David, Aaron T., J. Eli Asarian, and Frank K. Lake. "Wildfire Smoke Cools Summer River and Stream Water Temperatures." *Water Resources Research* 54, no. 10 (2018): 7273–90. https://doi.org/10.1029/2018WR022964.

Dawson, Graham. *Soldier Heroes: British Adventure, Empire and the Imagining of Masculinities*. Abingdon, Oxon: Taylor & Francis, 1994.

Deen, Alisha, Torri Estrada, Connor Everts, Sarah Farina, John Gibler, Amy Vanderwarker, Paola Ramos, and Cleo Woelfle-Erskine. *Thirsty for Justice: A People's Blueprint for California Water*. Oakland, CA: Environmental Justice Coalition for Water, 2005.

Delany, Samuel R. *The Motion of Light in Water: Sex and Science Fiction Writing in the East Village*. Minneapolis: University of Minnesota Press, 2004.

———. *Times Square Red, Times Square Blue*. New York: New York University Press, 2001.

Dellecave, J. *Nocturnal Beaver*. Endurance performance and accumulating sculpture. 2013. https://jdellecave.com/?p=147.

Deloria, Vine. *Red Earth, White Lies: Native Americans and the Myth of Scientific Fact*. Golden, CO: Fulcrum, 1997.

Denham, Diana, Mary Ann Rozance, Melanie Malone, and Erin Goodling. "Sustaining Future Environmental Educators: Building Critical Interdisciplinary Teaching Capacity among Graduate Students." *Journal of Environmental Studies and Sciences* (May 7, 2020). https://doi.org/10.1007/s13412-020-00611-y.

Dillon, Grace L., ed. *Walking the Clouds: An Anthology of Indigenous Science Fiction*. Tucson: University of Arizona Press, 2012.

Dillon, Lindsey. "Race, Waste, and Space: Brownfield Redevelopment and Environmental Justice at the Hunters Point Shipyard." *Antipode* 46, no. 5 (November 1, 2014): 1205–21. https://doi.org/10.1111/anti.12009.

———. "War's Remains: Slow Violence and the Urbanization of Military Bases in California." *Environmental Justice* 8, no. 1 (September 26, 2014): 1–5. https://doi.org/10.1089/env.2014.0014.

Diver, Sibyl Wentz. "Giving Back through Time: A Collaborative Timeline Approach to Researching Karuk Indigenous Land Management History." *Journal of Research Practice* 10, no. 2 (July 1, 2014): 18.

Dolman, Brock, and Kate Lundquist. *A Citizen's Guide to Protecting and Restoring Our Watersheds*. Occidental, CA: Occidental Arts and Ecology Center (OAEC), 2018.

Doroshow, Ceyenne, and Harron Walker. "The Black Trans Lives Matter March Was This Year's Pride." *Teen Vogue*. Accessed July 1, 2020. www.teenvogue.com/story/black-trans-lives-matter-march-ceyenne-doroshow.

D'Souza, Rohan. "Framing India's Hydraulic Crises." *Monthly Review: An Independent Socialist Magazine* 60, no. 3 (August 7, 2008): 112.

Duwamish River Cleanup Coalition. "Duwamish River Cleanup Coalition/Technical Advisory Group (DRCC/TAG)." Duwamish River Cleanup Coalition/Technical Advisory Group. Accessed March 29, 2019. http://duwamishcleanup.org.

Edelman, Elijah Adiv. *Trans Vitalities: Mapping Ethnographies of Trans Social and Political Coalitions*. New York: Routledge, 2020. https://doi.org/10.4324/9781351128025.

Ensor, Sarah. "Queer Fallout: Samuel R. Delany and the Ecology of Cruising." *Environmental Humanities* 9, no. 1 (May 1, 2017): 149–66. https://doi.org/10.1215/22011919-3829172.

Esselen Tribe of Monterey County. "Our History." Esselen Tribe of Monterey County. Accessed November 30, 2020. www.esselentribe.org/history.

Fabre, Jean Henri. *Fabre's Book of Insects*. Mineola, NY: Dover, 1998.

Federated Indians of Graton Rancheria. "History." Federated Indians of Graton Rancheria. Accessed February 6, 2021. https://gratonrancheria.com/culture/history/.

Ferguson, Roderick. "To Catch a Light-Filled Vision: American Studies and the Activation of Radical Traditions." Presented at the American Studies Association Annual Meeting, Atlanta, GA, November 9, 2019. www.youtube.com/watch?v=CMLBed3pDF8.

Flagg, Thomas A. "Balancing Conservation and Harvest Objectives: A Review of Considerations for the Management of Salmon Hatcheries in the U.S. Pacific Northwest." *North American Journal of Aquaculture* 77, no. 3 (July 3, 2015): 367–76. https://doi.org/10.1080/15222055.2015.1044058.

Foglia, L., J. Neumann, D. Tolley, S. Orloff, R. Snyder, and T. Harter. "Modeling Guides Groundwater Management in a Basin with River-Aquifer Interactions." *California Agriculture* 72, no. 1 (March 13, 2018): 84–95.

Foglia, Laura, Alison McNally, Courtney Hall, Lauren Ledesma, Ryan Hines, and Thomas Harter. "Scott Valley Integrated Hydrologic Model: Data Collection, Analysis, and Water Budget Final Report." UC Davis, 2013. groundwater.ucdavis.edu/files/165395.pdf.

Foglia, Laura, Alison McNally, and Thomas Harter. "Coupling a Spatiotemporally Distributed Soil Water Budget with Stream-Depletion Functions to

Inform Stakeholder-Driven Management of Groundwater-Dependent Ecosystems." *Water Resources Research* 49, no. 11 (November 1, 2013): 7292–7310. https://doi.org/10.1002/wrcr.20555.

Fortmann, Louise. *Participatory Research in Conservation and Rural Livelihoods: Doing Science Together.* Hoboken, NJ: John Wiley & Sons, 2009.Freeman, Elizabeth. "Queer Belongings: Kinship Theory and Queer Theory." In *A Companion to Lesbian, Gay, Bisexual, Transgender, and Queer Studies*, edited by George E. Haggerty and Molly McGarry, 293–314. Hoboken, NJ: John Wiley & Sons. https://doi.org/10.1002/9780470690864.ch15.

FUTURESTATES. "The 6th World." ITVS. Accessed June 4, 2018. www.youtube.com/watch?v=7f4Jmoy_iLk.

Gandy, Matthew. "Queer Ecology: Nature, Sexuality, and Heterotopic Alliances." *Environment and Planning D: Society and Space* 30, no. 4 (2012): 727–47. https://doi.org/10.1068/d10511.

Gardner, Karl, and Devin Clancy. "From Recognition to Decolonization: An Interview with Glen Coulthard." *Upping the Anti: A Journal of Theory and Action* 19 (2017). http://uppingtheanti.org/journal/article/19-from-recognition-to-decolonization.

Gillespie, Kathryn, and Patricia J. Lopez, eds. *Vulnerable Witness: The Politics of Grief in the Field.* Oakland: University of California Press, 2019.

Gómez-Barris, Macarena, and May Joseph. "Coloniality and Islands." *Shima: The International Journal of Research into Island Cultures* 13, no. 2 (September 26, 2019). https://doi.org/10.21463/shima.13.2.03.

Gordon, Avery F. "Some Thoughts on Haunting and Futurity." *Borderlands E-Journal* 10, no. 2 (2011): 1–21.

Gould, Deborah B. *Moving Politics: Emotion and ACT UP's Fight against AIDS.* Chicago: University of Chicago Press, 2009.

Great Lakes Indian Fish and Wildlife Commission. "Mercury Maps." Great Lakes Indian Fish and Wildlife Commission. Accessed October 20, 2020. www.glifwc.org/Mercury/.

Grossman, Zoltán. *Unlikely Alliances: Native Nations and White Communities Join to Defend Rural Lands.* Seattle: University of Washington Press, 2017.

Groves, Jason. *The Geological Unconscious.* New York: Fordham University Press, 2020. https://doi.org/10.2307/j.ctv1199127.

Halberstam, Jack, and Tavia Nyong'o. "Introduction: Theory in the Wild." *South Atlantic Quarterly* 117, no. 3 (July 1, 2018): 453–64. https://doi.org/10.1215/00382876-6942081.

Hammack, Lauren, L. Hulette, and Liza Prunuske. "Salmon Creek Integrated Coastal Watershed Management Plan." Gold Ridge Resource Conservation District, 2010.

Hammack, Lauren, Liza Prunuske, and Chris Choo. "Salmon Creek Estuary Study: Study Results and Enhancement Recommendations." Occidental, CA: Prunuske Chatham, 2006. www.bodeganet.com/landtrust/SC%20Estuary%20Final.pdf.

Hanak, Ellen, Jay R. Lund, Ariel Dinar, Richard Howitt, Jeffrey F. Mount, Peter B. Moyle, and Barton "Buzz" Thompson. *Managing California's Water: From Conflict to Reconciliation.* Public Policy Institute of CA, 2011.

Hannibal, Mary Ellen. "The Amah Mutsun and the Recovery of Traditional Ecological Knowledge." *Bay Nature.* Accessed November 30, 2020. https://baynature.org/article/rekindling-old-ways/.

Haraway, Donna. J. "Foreword: Companion Species, Mis-recognition, and Queer Worlding." In *Queering the Non/Human*, edited by Noreen Giffney and Myra J. Hird. Hampshire: Ashgate, 2008.

———. "A Manifesto for Cyborgs: Science, Technology, and Socialist Feminism in the 1980s." *Australian Feminist Studies* 2, no. 4 (1987): 1–42.

———. "Situated Knowledges: The Science Question in Feminism and the Privilege of Partial Perspective." *Feminist Studies* 14, no. 3 (1988): 575–99.

———. *When Species Meet.* Minneapolis: University of Minnesota Press, 2008.

Harding, Sandra. *Sciences from Below: Feminisms, Postcolonialities, and Modernities.* Durham, NC: Duke University Press, 2008.

———. "'Strong Objectivity': A Response to the New Objectivity Question." *Synthese* 104, no. 3 (1995): 331–49.

Haritaworn, Jinthana. "Decolonizing the Non/Human." *GLQ: A Journal of Lesbian and Gay Studies* 21, no. 2 (May 9, 2015): 210–13.

Harling, Will. "One Beaver at a Time." Orleans: Mid Klamath Watershed Council, n.d.

Harney, Stefano, and Fred Moten. *The Undercommons: Fugitive Planning & Black Study.* Autonomedia, 2013.

Harris, Preston. "2013–15 Three Year Summary." Etna, CA: Scott River Water Trust, n.d. https://drive.google.com/file/d/0B5MBuiH39wUqaE05TDJ1 TmlBZEt3akJ5aFBEQ1pqc3JoazhB/view.

Hayward, Eva. "Lessons from a Starfish." In *Queering the Non/Human*, edited by Noreen Giffney and Myra J. Hird. Hampshire: Ashgate, 2008.

Herring, Scott. *Another Country: Queer Anti-Urbanism.* New York: NYU Press, 2010.

Holifield, Ryan. "Environmental Justice as Recognition and Participation in Risk Assessment: Negotiating and Translating Health Risk at a Superfund Site in Indian Country." *Annals of the Association of American Geographers* 102, no. 3 (May 1, 2012): 591–613. https://doi.org/10.1080/00045608.2011.641892.

Hustak, Carla, and Natasha Myers. "Involutionary Momentum: Affective Ecologies and the Sciences of Plant/Insect Encounters." *Differences* 23, no. 3 (January 1, 2012): 74–118. https://doi.org/10.1215/10407391-1892907.

Jasanoff, Sheila, and Sang-Hyun Kim. "Sociotechnical Imaginaries and National Energy Policies." *Science as Culture* 22, no. 2 (June 1, 2013): 189–96. https://doi.org/10.1080/09505431.2013.786990.

Johannson, Ariele. "Warner Springs Ranch Comes Full Circle with Pala Purchase." *East County Magazine.* Accessed November 30, 2020. www

.eastcountymagazine.org/warner-springs-ranch-comes-full-circle-pala-purchase.

Kafer, Alison. *Feminist, Queer, Crip.* Bloomington: Indiana University Press, 2013.

Karuk Tribe. "Land Management." Karuk Tribe. Accessed November 30, 2020. www.karuk.us/index.php/departments/land-management.

Katz, J., P. B. Moyle, R. M. Quiñones, J. Israel, and S. Purdy. "Impending Extinction of Salmon, Steelhead, and Trout (*Salmonidae*) in California." *Environmental Biology of Fishes* 96, no. 10–11 (2012): 1–18.

Kauanui, J. Kēhaulani. "'A Structure, Not an Event': Settler Colonialism and Enduring Indigeneity,'" *Lateral* (June 1, 2016). https://csalateral.org/issue/5-1/forum-alt-humanities-settler-colonialism-enduring-indigeneity-kauanui/.

Khan, Naveeda. "At Play with the Giants: Between the Patchy Anthropocene and Romantic Geology." *Current Anthropology* 60, no. S20 (March 22, 2019): S333–41. https://doi.org/10.1086/702756.

Kim, Esther G. "Bring on the Yuppies and the Guppies! Green Gentrification, Environmental Justice, and the Politics of Place in Frogtown, L.A." In *Just Green Enough*, edited by Winifred Curran and Trina Hamilton. Abingdon, Oxon: Routledge, 2017.

Kimmerer, Robin Wall. *Braiding Sweetgrass: Indigenous Wisdom, Scientific Knowledge and the Teachings of Plants.* Minneapolis, MN: Milkweed, 2015.

———. "The Covenant of Reciprocity." In *The Wiley Blackwell Companion to Religion and Ecology*, edited by John Hart, 368–82. Hoboken, NJ: Wiley-Blackwell, 2017.

King, Thomas. *First Salmon: The Klamath Cultural Riverscape and the Klamath River Hydroelectric Project.* Chiloquin, OR: Klamath River Intertribal Fish and Water Commission, 2004. http://resighinirancheria.com/Documents/Klamath_Riverscape_King_2004.pdf.

Kiparsky, Michael, Anita Milman, and Sebastian Vicuña. "Climate and Water: Knowledge of Impacts to Action on Adaptation." *Annual Review of Environment and Resources* 37, no. 1 (2012): 163–94. https://doi.org/10.1146/annurev-environ-050311-093931.

Kirchhoff, Christine J., Maria Carmen Lemos, and Suraje Dessai. "Actionable Knowledge for Environmental Decision Making: Broadening the Usability of Climate Science." *Annual Review of Environment and Resources* 38, no. 1 (2013): 393–414. https://doi.org/10.1146/annurev-environ-022112-112828.

Kirksey, S. Eben, Nicholas Shapiro, and Maria Brodine. "Hope in Blasted Landscapes." *Social Science Information* 52, no. 2 (June 1, 2013): 228–56. https://doi.org/10.1177/0539018413479468.

Klamath Riverkeeper. "Comments to the Scott Valley Groundwater Study Plan," October 2007. http://groundwater.ucdavis.edu/files/136428.pdf.

Koenigsberg, Sarah. *The Beaver Believers*, 2018. www.thebeaverbelievers.com.

Koran, Mario. "Northern California Esselen Tribe Regains Ancestral Land after 250 Years." *Guardian.* July 28, 2020. www.theguardian.com/us-news/2020/jul/28/northern-california-esselen-tribe-regains-land-250-years.

Korinek, Valerie. *Prairie Fairies: A History of Queer Communities and People in Western Canada, 1930–1985.* Toronto: University of Toronto Press, 2018.

Kormann, Carolyn. "How Carbon Trading Became a Way of Life for California's Yurok Tribe." *New Yorker.* Accessed November 30, 2020. www.newyorker.com/news/dispatch/how-carbon-trading-became-a-way-of-life-for-californias-yurok-tribe.

Krupar, Shiloh R. *Hot Spotter's Report: Military Fables of Toxic Waste.* Minneapolis: University of Minnesota Press, 2013.

Larsen, Laurel G., and Cleo Woelfle-Erskine. "Groundwater Is Key to Salmonid Persistence and Recruitment in Intermittent Mediterranean-Climate Streams." *Water Resources Research,* 2018.

Lave, Rebecca, Christine Biermann, and Stuart N. Lane. "Introducing Critical Physical Geography." In *The Palgrave Handbook of Critical Physical Geography,* edited by Rebecca Lave, Christine Biermann, and Stuart N. Lane, 3–22. London: Palgrave Macmillan, 2018.

Li, Tania Murray. *The Will to Improve: Governmentality, Development, and the Practice of Politics.* Durham, NC: Duke University Press, 2007. https://doi.org/10.2307/j.ctv11smt9s.

Liboiron, Max. "R-Words: Refusing Research." *Action-Based Research Methods* (blog). March 28, 2016. https://activistresearchmethods.wordpress.com/2016/03/28/r-words-refusing-research.

Lin, Maya. *What Is Missing?* Accessed April 8, 2019. www.whatismissing.org/.

Linton, Jamie. *What Is Water? The History of a Modern Abstraction.* Vancouver, BC: UBC Press, 2010.

Linton, Jamie, and Jessica Budds. "The Hydrosocial Cycle: Defining and Mobilizing a Relational-Dialectical Approach to Water." *Geoforum* 57 (2013): 170–80. https://doi.org/10.1016/j.geoforum.2013.10.008.

Malone, Melanie. "Using Critical Physical Geography to Map the Unintended Consequences of Conservation Management Programs." Portland State University, 2017. https://doi.org/10.15760/etd.5523.

Marks-Block, Tony, Frank K. Lake, and Lisa M. Curran. "Effects of Understory Fire Management Treatments on California Hazelnut, an Ecocultural Resource of the Karuk and Yurok Indians in the Pacific Northwest." *Forest Ecology and Management* 450 (October 15, 2019): 117517. https://doi.org/10.1016/j.foreco.2019.117517.

Martinez, Dennis. "Protected Areas, Indigenous Peoples, and the Western Idea of Nature." *Ecological Restoration* 21, no. 4 (2003): 247–50.

Martinez, Donna. "Land Returned to Pomo Tribe Forms Kashia Coastal Reserve." Sea Ranch Abalone Bay. October 29, 2015. https://searanchabalonebay.com/news/kashia-coastal-reserve-700-acres-of-ranch-land-returned-to-pomo-tribe.

McHugh, Maureen F. *After the Apocalypse: Stories*. Easthampton, MA: Small Beer, 2011.

McKittrick, Katherine. *Dear Science and Other Stories*. Durham, NC: Duke University Press, 2021.

Mid-Columbia Fisheries Enhancement Group. "Yakima Beaver Project." Mid-Columbia Fisheries Enhancement Group. Accessed September 20, 2020. http://midcolumbiafisheries.org/restoration/beaver-restoration/beaver-reintroduction/.

Ming-Yi, Wu. *The Man with the Compound Eyes: A Novel*. New York: Pantheon, 2014.

Morgensen, Scott Lauria. *Spaces between Us: Queer Settler Colonialism and Indigenous Decolonization*. Minneapolis: University of Minnesota Press, 2011.

Muñoz, José Esteban. "After Jack: Queer Failure, Queer Virtuosity." In *Cruising Utopia: The Then and There of Queer Futurity*, 169–84. New York: NYU Press, 2009.

———. "The Brown Commons: The Sense of Wildness." Presented at the *Journal of Narrative Theory* dialogue Queer Commons, Ypsilanti, MI, March 25, 2013. www.youtube.com/watch?v=F-YInUlXgO4.

———. *Cruising Utopia: The Then and There of Queer Futurity*. New York: NYU Press, 2009.

———. "Ephemera as Evidence: Introductory Notes to Queer Acts." *Women & Performance: A Journal of Feminist Theory* 8, no. 2 (January 1, 1996): 5–16. https://doi.org/10.1080/07407709608571228.

———. "Feeling Brown." Presented at the University of Maryland, College Park, MD, March 3, 2002. https://vimeo.com/282566570.

———. "Feeling Brown, Feeling Down: Latina Affect, the Performativity of Race, and the Depressive Position." *Signs* 31, no. 3 (2006): 675–88. https://doi.org/10.1086/499080.

———. "Gestura, Ephemera, and Queer Feeling: Approaching Kevin Aviance." In *Cruising Utopia: The Then and There of Queer Futurity*, 65–82. New York: NYU Press, 2009.

———. "'Gimme Gimme This . . . Gimme Gimme That': Annihilation and Innovation in the Punk Rock Commons." *Social Text* 31, no. 3 (116) (September 1, 2013): 95–110. https://doi.org/10.1215/01642472-2152855.

———. "Keynote: The Wildness of the Commons." Presented at We Who Feel Differently: A Symposium, New Museum, New York, NY, May 5, 2014. https://wewhofeeldifferently.info/ephemera.php#Symposium.

———. "Preface: Fragment from the *Sense of Brown* Manuscript." *GLQ: A Journal of Lesbian and Gay Studies* 24, no. 4 (October 25, 2018): 395–97.

———. *The Sense of Brown*. Edited by Joshua Chambers-Letson and Tavia Nyong'o. Durham, NC: Duke University Press, 2020.

———. "The Wildness of the Punk Rock Commons." *South Atlantic Quarterly* 117, no. 3 (July 1, 2018): 653–58. https://doi.org/10.1215/00382876-6942219.

Muñoz, José Esteban, Jinthana Haritaworn, Myra Hird, Zakiyyah Iman Jackson, Jasbir K. Puar, Eileen Joy, Uri McMillan, et al. "Theorizing Queer Inhumanisms." *GLQ: A Journal of Lesbian and Gay Studies* 21, no. 2–3 (January 1, 2015): 209–48. https://doi.org/10.1215/10642684-2843323.

Murphy, Michelle. "Unsettling Care: Troubling Transnational Itineraries of Care in Feminist Health Practices." *Social Studies of Science* 45, no. 5 (October 1, 2015): 717–37. https://doi.org/10.1177/0306312715589136.

Nadasdy, P. *Hunters and Bureaucrats: Power, Knowledge, and Aboriginal-State Relations in the Southwest Yukon*. Vancouver, BC: UBC Press, 2004.

Nancy, Jean-Luc. *Being Singular Plural*. Translated by Robert Richardson and Anne O'Byrne. Stanford, CA: Stanford University Press, 2000.

National Marine Fisheries Service. *Final Recovery Plan for Central California Coast Coho Salmon Evolutionarily Significant Unit*. Santa Rosa, CA: National Marine Fisheries Service, Southwest Region, September 2012.

———. *Southern Oregon Northern California Coast Coho Salmon Recovery Plan*. Santa Rosa, CA: National Marine Fisheries Service, Southwest Region, January 1, 2014. www.westcoast.fisheries.noaa.gov/protected_species/salmon_steelhead/recovery_planning_and_implementation/southern_oregon_northern_california_coast/SONCC_recovery_plan.html.

Neimanis, Astrida, Greg Garrard, and Richard Kerridge. *Bodies of Water: Posthuman Feminist Phenomenology*. London: Bloomsbury Academic, 2017.

Nichols, A. L., A. D. Willis, C. A. Jeffres, and M. L. Deas. "Water Temperature Patterns below Large Groundwater Springs: Management Implications for Coho Salmon in the Shasta River, California." *River Research and Applications* 30, no. 4 (May 1, 2014): 442–55. https://doi.org/10.1002/rra.2655.

Nokes, Kelly. "An Opportunity to Protect—Analyzing Fish Consumption, Environmental Justice, and Water Quality Standards Rulemaking in Washington State." *Vermont Journal of Environmental Law* 16, no. 2 (Fall 2014): 323–62.

Norgaard, Kari Marie. "The Politics of Fire and the Social Impacts of Fire Exclusion on the Klamath." *Humboldt Journal of Social Relations* 36 (2014): 25.

Norgaard, Kari Marie, Kristen Vinyeta, Leaf Hillman, Bill Tripp, and Frank Lake. *Karuk Tribe Climate Vulnerability Assessment: Assessing Vulnerabilities from the Increased Frequency of High Severity Fire*. Orleans, CA: Karuk Tribe Department of Natural Resources, 2016.

Nyong'o, Tavia. "Brown Punk: Kalup Linzy's Musical Anticipations." *TDR/The Drama Review* 54, no. 3 (August 25, 2010): 71–86. https://doi.org/10.1162/DRAM_a_00005.

Okorafor-Mbachu, Nnedi. *Zahrah the Windseeker*. Reprint, Boston: HMH Books for Young Readers, 2008.

Ostrom, Elinor. *Governing the Commons: The Evolution of Institutions for Collective Action*. Cambridge, MA: Cambridge University Press, 1990.

Parreñas, Juno Salazar. *Decolonizing Extinction: The Work of Care in Orangutan Rehabilitation*. Durham, NC: Duke University Press, 2018.

Peet, Richard, and Michael Watts. *Liberation Ecologies: Environment, Development, Social Movements.* London: Routledge, 1996.

Pollock, Michael M., Timothy J. Beechie, and Chris E. Jordan. "Geomorphic Changes Upstream of Beaver Dams in Bridge Creek, an Incised Stream Channel in the Interior Columbia River Basin, Eastern Oregon." *Earth Surface Processes and Landforms* 32, no. 8 (2007): 1174–85.

Pollock, Michael M., Timothy J. Beechie, Joseph M. Wheaton, Chris E. Jordan, Nick Bouwes, Nicholas Weber, and Carol Volk. "Using Beaver Dams to Restore Incised Stream Ecosystems." *BioScience* 64, no. 4 (April 1, 2014): 279–90. https://doi.org/10.1093/biosci/biu036.

Power, Mary E., Michael S. Parker, and William E. Dietrich. "Seasonal Reassembly of a River Food Web: Floods, Droughts, and Impacts of Fish." *Ecological Monographs* 78, no. 2 (May 2008): 263–82. https://doi.org/10.1890/06-0902.1.

Power, Mary E., David Tilman, James A. Estes, Bruce A. Menge, William J. Bond, L. Scott Mills, Gretchen Daily, Juan Carlos Castilla, Jane Lubchenco, and Robert T. Paine. "Challenges in the Quest for Keystones: Identifying Keystone Species Is Difficult—but Essential to Understanding How Loss of Species Will Affect Ecosystems." *BioScience* 46, no. 8 (September 1, 1996): 609–20. https://doi.org/10.2307/1312990.

Prunuske Chatham Inc., Virginia Porter, and OAEC WATER Institute. *Salmon Creek Water Conservation Plan.* Occidental, CA: Occidental Arts and Ecology Center, 2010. www.salmoncreekwater.org/water-conservation-plan.html.

psmfcvideo. *Beavers on Working Lands Featuring Landowners Betsy and Michael Stapleton.* Accessed February 15, 2019. www.youtube.com/watch?v=eulioYwyr2c.

Puar, Jasbir K. "'I Would Rather Be a Cyborg than a Goddess': Becoming-Intersectional in Assemblage Theory." *PhiloSOPHIA* 2, no. 1 (2012): 49–66.

Rechy, John. *City of Night.* Anniv. ed. New York: Grove, 2013.

Reisner, Marc. *Cadillac Desert: The American West and Its Disappearing Water.* New York: Penguin Books, 1993.

Robinson, Kim Stanley. 2312. London: Orbit Books, 2013.

———. *Antarctica: A Novel.* Reissue, New York: Bantam, 1999.

Rose, Deborah Bird. "Double Death." *Deborah Bird Rose* (blog). Accessed April 28, 2015. http://deborahbirdrose.com/144-2/.

Samuels, Ellen. "Six Ways of Looking at Crip Time." *Disability Studies Quarterly* 37, no. 3 (August 31, 2017). https://doi.org/10.18061/dsq.v37i3.5824.

Sarna-Wojcicki, Daniel Reid. "Scales of Sovereignty: The Search for Watershed Democracy in the Klamath Basin." EScholarship. January 1, 2015. http://escholarship.org/uc/item/3tk5r3w2.

Sarna-Wojcicki, Daniel, Jennifer Sowerwine, and Lisa Hillman. "Decentring Watersheds and Decolonising Watershed Governance: Towards an Ecocultural Politics of Scale in the Klamath Basin." *Water Alternatives* 12, no. 1 (2019): 26.

Sarna-Wojcicki, Daniel, and Cleo Woelfle-Erskine. "The Manifest Reversals of Multi-species Collaborative Watershed Restoration." Presented at the Dimensions of Political Ecology conference, Lexington, KY, March 1, 2014.

Schell, Christopher J., Karen Dyson, Tracy L. Fuentes, Simone Des Roches, Nyeema C. Harris, Danica Sterud Miller, Cleo A. Woelfle-Erskine, and Max R. Lambert. "The Ecological and Evolutionary Consequences of Systemic Racism in Urban Environments." *Science*. August 13, 2020. https://doi.org/10.1126/science.aay4497.

Schell, Christopher J., Cylita Guy, Delia S. Shelton, Shane C. Campbell-Staton, Briana A. Sealey, Danielle N. Lee, and Nyeema C. Harris. "Recreating Wakanda by Promoting Black Excellence in Ecology and Evolution." *Nature Ecology & Evolution* 4, no. 10 (October 2020): 1285–87. https://doi.org/10.1038/s41559-020-1266-7.

Schrader, Astrid. "Responding to *Pfiesteria piscicida* (the Fish Killer): Phantomatic Ontologies, Indeterminacy, and Responsibility in Toxic Microbiology." *Social Studies of Science* 40, no. 2 (February 15, 2010). http://journals.sagepub.com/doi/10.1177/0306312709344902.

Schuller, Kyla. *The Biopolitics of Feeling: Race, Sex, and Science in the Nineteenth Century*. Illus. ed. Durham, NC: Duke University Press, 2017.

Scott, James C. *Seeing Like a State: How Certain Schemes to Improve the Human Condition Have Failed*. Yale University Press, 1998.

Scott River Watershed Council. *Restoring Priority Coho Habitat in the Scott River Watershed Modeling and Planning Report*. Etna, CA: Scott River Watershed Council, 2018. www.scottriverwatershedcouncil.com.

Scott Valley Groundwater Action Committee. *Voluntary Groundwater Management & Enhancement Plan for Scott Valley Advisory Committee (Approved 10-22-12)*. Etna, CA: Scott Valley Groundwater Action Committee, October 22, 2012.

Sedgwick, Eve Kosofsky. "A Poem Is Being Written." *Representations*, no. 17 (January 1, 1987): 110–43. https://doi.org/10.2307/3043795.

———. *Tendencies*. Durham, NC: Duke University Press, 1993.

———. "White Glasses." In *Tendencies*, 252–66. Durham, NC: Duke University Press, 1993.

Simpson, Audra. *Mohawk Interruptus: Political Life across the Borders of Settler States*. Durham, NC: Duke University Press, 2014.

———. "Settlement's Secret." *Cultural Anthropology* 26, no. 2 (May 1, 2011): 205–17. https://doi.org/10.1111/j.1548-1360.2011.01095.x.

Simpson, Leanne Betasamosake. *As We Have Always Done: Indigenous Freedom through Radical Resistance*. 3rd ed. Minneapolis: University of Minnesota Press, 2017.

———. "Indigenous Resurgence and Co-resistance." *Critical Ethnic Studies* 2, no. 2 (2016): 19–34. https://doi.org/10.5749/jcritethnstud.2.2.0019.

———. *This Accident of Being Lost: Songs and Stories*. Toronto, ON: House of Anansi, 2017.

Siskiyou Co. Farm Bureau v. Dept. Fish & Wildlife, No. SCCVCV1100418 (Siskiyou County Superior Court 06042015).

Sogorea Te Land Trust. "Shuumi Land Tax." Accessed November 30, 2020. https://sogoreate-landtrust.org/shuumi-land-tax/.

Spicer, Jack. "Improvisations on a Sentence by Poe." In *A Book of Music*. San Francisco: White Rabbit, 1969.

Stanford, Jack A., and J. V. Ward. "An Ecosystem Perspective of Alluvial Rivers: Connectivity and the Hyporheic Corridor." *Journal of the North American Benthological Society* 12, no. 1 (March 1, 1993): 48–60. https://doi.org/10.2307/1467685.

Stanley, Eric. "The Affective Commons: Gay Shame, Queer Hate, and Other Collective Feelings." *GLQ: A Journal of Lesbian and Gay Studies* 24, no. 4 (October 1, 2018): 489–508. https://doi.org/10.1215/10642684-6957800.

Stryker, Susan. "(De)Subjugated Knowledges: An Introduction to Transgender Studies." In *The Transgender Studies Reader*, edited by Susan Stryker and Stephen Whittle, 1–17. New York: Routledge, 2006.

Subramaniam, Banu. *Ghost Stories for Darwin: The Science of Variation and the Politics of Diversity*. Urbana: University of Illinois Press, 2014.

Subramaniam, Banu, and Angela Willey. "Introduction: Feminism's Sciences." *Catalyst: Feminism, Theory, Technoscience* 3, no. 1 (2017): 1–23. https://catalystjournal.org/index.php/catalyst/article/view/28784.

Sycamore, Mattilda Bernstein, ed. *That's Revolting! Queer Strategies for Resisting Assimilation*. Rev. ed. Brooklyn: Soft Skull, 2008.

Sze, Julie, Jonathan London, Fraser Shilling, Gerardo Gambirazzio, Trina Filan, and Mary Cadenasso. "Defining and Contesting Environmental Justice: Socio-natures and the Politics of Scale in the Delta." *Antipode* 41, no. 4 (September 2009): 807–43. https://doi.org/10.1111/j.1467-8330.2009.00698.x.

Tafoya, Terry. "Finding Harmony: Balancing Traditional Values with Western Science in Therapy." *Canadian Journal of Native Education* 21 (1995): suppl. p7–27.

TallBear, Kimberly. *The Critical Polyamorist: Polyamory, Indigeneity, and Cultural Politics in the US and Canada* (blog). Accessed March 28, 2019. www.criticalpolyamorist.com.

———. "Genomic Articulations of Indigeneity." *Social Studies of Science* 43, no. 4 (2013): 509–33.

———. *Making Love and Relations beyond Settler Sexualities*. Accessed November 19, 2017. http://indigenoussts.com/event/making-love-and-relations-beyond-settler-sexualities/.

———. "Standing with and Speaking as Faith: A Feminist-Indigenous Approach to Inquiry." *Journal of Research Practice* 10, no. 2 (July 1, 2014): Article N17.

———. "Uppsala 3rd Supradisciplinary Feminist Technoscience Symposium: Feminist and Indigenous Intersections and Approaches to Technoscience."

Kim TallBear: Indigeneity & Technoscience (blog). Accessed September 29, 2014. www.kimtallbear.com.

Taylor, Sonya Renee. *The Body Is Not an Apology: The Power of Radical Self-Love.* Oakland, CA: Berrett-Koehler, 2018.

Thrush, Coll. "City of the Changers." *Pacific Historical Review* 75, no. 1 (2006): 89–117. https://doi.org/10.1525/phr.2006.75.1.89.

———. *Native Seattle: Histories from the Crossing-Over Place.* Seattle: University of Washington Press, 2008.

Todd, Zoe. "Fish, Kin and Hope: Tending to Water Violations in Amiskwaciwâskahikan and Treaty Six Territory." *Afterall: A Journal of Art, Context and Enquiry* 43 (March 1, 2017): 102–7. https://doi.org/10.1086/692559.

———. "Fish Pluralities: Human-Animal Relations and Sites of Engagement in Paulatuuq, Arctic Canada." *Études/Inuit/Studies* 38, no. 1–2 (2014): 217. https://doi.org/10.7202/1028861ar.

———. "From a Fishy Place: Examining Canadian State Law Applied in the Daniels Decision from the Perspective of Métis Legal Orders." *TOPIA: Canadian Journal of Cultural Studies* 36 (2016): 43–57. https://topia.journals.yorku.ca/index.php/topia/article/view/40398.

Tomblin, David C. "The Ecological Restoration Movement: Diverse Cultures of Practice and Place." *Organization & Environment* 22, no. 2 (June 2009): 185–207. https://doi.org/10.1177/1086026609338165.

Trosper, Ronald. 2003. "Resilience in Pre-contact Pacific Northwest Social Ecological Systems." *Conservation Ecology* 7 (3). https://doi.org/10.5751/ES-00551-070306.

Tsang, Wu. *Wildness*, 2012. www.imdb.com/title/tt1758837/.

Tsing, Anna, for the Matsutake Worlds Research Group. "Arts of Inclusion, or How to Love a Mushroom." *Australian Humanities Review*, no. 50 (2011): 191–205. http://epress.anu.edu.au/apps/bookworm/view/Australian+Humanities+Review+-+Issue+50,+2011/5451/ch01.xhtml.

Tsing, Anna Lowenhaupt. *Friction: An Ethnography of Global Connection.* Princeton, NJ: Princeton University Press, 2004.

———. *The Mushroom at the End of the World: On the Possibility of Life in Capitalist Ruins.* Princeton, NJ: Princeton University Press, 2015.

Tuck, Eve. "Breaking up with Deleuze: Desire and Valuing the Irreconcilable." *International Journal of Qualitative Studies in Education* 23, no. 5 (September 1, 2010): 635–50. https://doi.org/10.1080/09518398.2010.500633.

Tuck, Eve, and K. Wayne Yang. "Decolonization Is Not a Metaphor." *Decolonization: Indigeneity, Education & Society* 1, no. 1 (September 8, 2012): 1–40.

———. "R-Words: Refusing Research." In *Humanizing Research: Decolonizing Qualitative Inquiry with Youth and Communities*, edited by D. Paris and M. T. Winn, 223–47. Thousand Oaks, CA: SAGE, 2014.

Tulalip Tribes Natural Resources. "Beaver." Tulalip Tribes Natural Resources. Accessed September 20, 2020. https://nr.tulaliptribes.com/Programs/Wildlife/Beaver.

Turnbull, David. *Masons, Tricksters and Cartographers: Comparative Studies in the Sociology of Scientific and Indigenous Knowledge*. Amsterdam: Taylor & Francis, 2000.

UC Cooperative Extension. "Russian River Coho Salmon Recovery Program." UCCE Sonoma County. 2015. http://cesonoma.ucanr.edu/Marine_Science-Sea_Grant585/Captive_Broodstock_Recovery.

Van Kirk, Robert W., and Seth W. Naiman. "Relative Effects of Climate and Water Use on Base-Flow Trends in the Lower Klamath Basin." *JAWRA Journal of the American Water Resources Association* 44, no. 4 (August 1, 2008): 1035–52. https://doi.org/10.1111/j.1752-1688.2008.00212.x.

Water Education Foundation. "Overdraft." Water Education Foundation. Accessed December 1, 2020. www.watereducation.org/aquapedia/overdraft.

Waterlines. "The Waterlines Project Map." Accessed November 23, 2020. Burke Museum. www.burkemuseum.org/static/waterlines/project_map.html.

Weir, Jessica K. *Murray River Country: An Ecological Dialogue with Traditional Owners*. Canberra: Aboriginal Studies Press, 2009.

Wells, Harry Laurenz. *History of Siskiyou County, California: Illustrated with Views of Residences, Business Buildings and Natural Scenery, and Containing Portraits and Biographies of Its Leading Citizens and Pioneers*. Oakland, CA: D. J. Stewart, 1881.

Wheaton, Joseph M., Stephen E. Darby, and David A. Sear. "The Scope of Uncertainties in River Restoration." In *River Restoration*, edited by Stephen Darby and David Sear, 21–39. Hoboken, NJ: John Wiley & Sons, 2008. https://doi.org/10.1002/9780470867082.ch3.

White, Richard. *The Organic Machine: The Remaking of the Columbia River*. New York: Hill and Wang, 1996.

Whyte, Kyle Powys. "Food Sovereignty, Justice and Indigenous Peoples: An Essay on Settler Colonialism and Collective Continuance." In *The Oxford Handbook of Food Ethics*, edited by Anne Barnhill, Tyler Doggett, and Mark Budolfson, 345–66. New York: Oxford University Press, 2018.

———. *Indigeneity and US Settler Colonialism*. Vol. 1, edited by Naomi Zack. New York: Oxford University Press, 2017. https://doi.org/10.1093/oxfordhb/9780190236953.013.51.

———. "Our Ancestors' Dystopia Now: Indigenous Conservation and the Anthropocene." In *Routledge Companion to the Environmental Humanities*, edited by Ursula K. Heise, Jon Christensen, and Michelle Niemann, 206–15. London: Routledge, 2017. https://papers.ssrn.com/abstract=2770047.

Whyte, Kyle, Chris Caldwell, and Marie Schaefer. "Indigenous Lessons about Sustainability Are Not Just for 'All Humanity.'" In *Sustainability*, edited by Julie Sze, 149–79. New York: NYU Press, 2018. https://doi.org/10.18574/nyu/9781479894567.003.0007.

Wiegman, Robyn. "The Times We're In: Queer Feminist Criticism and the Reparative 'Turn.'" *Feminist Theory* 15, no. 1 (April 1, 2014): 4–25. https://doi.org/10.1177/1464700113513081a.

Wilkinson, Charles. *Messages from Frank's Landing: A Story of Salmon, Treaties, and the Indian Way*. Seattle: University of Washington Press, 2006.

Willette, Mirranda, Kari Norgaard, and Ron Reed. "You Got to Have Fish: Families, Environmental Decline and Cultural Reproduction." *Families, Relationships and Societies* 5, no. 3 (November 18, 2016): 375–92. https://doi.org/10.1332/204674316X14758424912055.

Willey, Angela. "Biopossibility: A Queer Feminist Materialist Science Studies Manifesto, with Special Reference to the Question of Monogamous Behavior." *Signs: Journal of Women in Culture and Society* 41, no. 3 (February 16, 2016): 553–77. https://doi.org/10.1086/684238.

Wilson, Shawn. *Research Is Ceremony: Indigenous Research Methods*. Black Point, NS: Fernwood Publishing, 2009.

Woelfle-Erskine, Cleo. "Collaborative Approaches to Flow Restoration in Intermittent Salmon-Bearing Streams: Salmon Creek, CA, USA." *Water* 9, no. 3 (March 14, 2017): 217. https://doi.org/10.3390/w9030217.

———. "Emerging Cultural Waterscapes in California Cities Connect Rain to Taps and Drains to Gardens." In *Sustainable Water: Challenges and Solutions from California*, edited by Allison Lassiter, 317–41. Berkeley, CA: University of California Press, 2017.

———. "Fishy Pleasures: Unsettling Fish Hatching and Fish Catching on Pacific Frontiers." *Imaginations: Journal of Cross-Cultural Image Studies* 10, no. 1 (July 25, 2019): 325–52. https://doi.org/10.17742/IMAGE.CR.10.1.11.

———. "Logjams, Check Dams, and the Gift of Foresight: Restoration on Native Land." In *Dam Nation: Dispatches from the Water Underground*, edited by Cleo Woelfle-Erskine, July Oskar Cole, Laura Allen, and Annie Danger, 246–70. Illus. ed. New York: Soft Skull Press, 2007.

———. "Rain Tanks, Springs, and Broken Pipes as Emerging Water Commons along Salmon Creek, CA, USA." *ACME: An International E-Journal for Critical Geographies* 14, no. 3 (September 26, 2015): 735–50.

———. "Thinking with Salmon about Rain Tanks: Commons as Intra-Actions." *Local Environment* 20, no. 5 (May 4, 2015): 581–99. https://doi.org/10.1080/13549839.2014.969212.

———. "Who Needs Dams?" In *Dam Nation: Dispatches from the Water Underground*, edited by Cleo Woelfle-Erskine, July Oskar Cole, Laura Allen, and Annie Danger, 13–42. Illus. ed. Brooklyn: Soft Skull, 2007.

Woelfle-Erskine, Cleo, and July Oskar Cole. "Transfiguring the Anthropocene: Stochastic Re-imaginings of Human Beaver Worlds." *Transgender Studies Quarterly* 2, no. 2 (2015): 297–316. https://doi.org/10.1215/23289252-2867625.

Woelfle-Erskine, Cleo, July Oskar Cole, Laura Allen, and Annie Danger. *Dam Nation: Dispatches from the Water Underground*. Illus. ed. New York: Soft Skull Press, 2007.

Woelfle-Erskine, Cleo, Laurel G. Larsen, and Stephanie M. Carlson. "Abiotic Habitat Thresholds for Salmonid Over-Summer Survival in Intermittent Streams." *Ecosphere* 8, no. 2 (February 1, 2017): 1–23. https://doi.org/10.1002/ecs2.1645.

Woelfle-Erskine, Cleo, and Daniel Sarna-Wojcicki. "Dam 'Em All: Beaver Believers, Beaver Deceivers, and Other Watershed Entanglements." UC Berkeley, 2013. www.youtube.com/watch?v=9IcMy8W_68M&feature=youtube_gdata_player.

———. "Hyporheic Imaginaries: How Beaver Collaborations Remix Patterns of Wet and Dry." In *Amphibious Anthropologies: Human Lives between Wet and Dry*, in review.

Wojnarowicz, David. *Close to the Knives: A Memoir of Disintegration*. New York: Vintage, 1991.

———. *The Waterfront Journals*. New York: Grove, 1997.

Worster, Donald. "Hydraulic Society in California: An Ecological Interpretation." *Agricultural History* 56, no. 3 (July 1, 1982): 503–15.

———. *Rivers of Empire: Water, Aridity, and the Growth of the American West*. New York: Oxford University Press, 1985.

Worth a Dam. Worth A Dam—The Martinez Beavers. Accessed March 28, 2019. www.martinezbeavers.org/wordpress/.

Yoshiyama, Ronald M., and Frank W. Fisher. "Long Time Past: Baird Station and the McCloud Wintu." *Fisheries* 26, no. 3 (March 1, 2001): 6–22. https://doi.org/10.1577/1548-8446(2001)026<0006:LTPBSA>2.0.CO;2.

Index

Italicized page numbers refer to figures

@500queerScientists, 142
#CuteOff, 33, 86, 88

ACT UP, 100, 147, 155
Adams, Ansel, 208
Agard-Jones, Vanessa, 202
agential realism, 44
Ahmed, Sara, 155
Aizura, Aren, 80
alder trees, xiv, 10, 40, 87, 113
alfalfa, xiv, xvi, 40, 45, 58–59, 66, 72
algae, 10, 12, 19, 36, 219, 221
American Geophysical Meeting: GayGU, 142
American School of Ethnology, 154
American Studies Association, 168, 182
the amphibious, xvi–xvii, 204, 221–22
Anderson, Carl, 52, 54–56
Anderson, Rob, 170
animacy, 45, 92, 101, 136, 191, 244n50
Anishinaabe Peoples, 213
anthropocene, 107, 120, 122, 135, 204
anti-Blackness, 22, 25, 29, 117, 210, 228n4
antiracism, 83, 218, 249n42
Anzaldúa, Gloria, 83, 179, 202

apparatus (Barad), 32, 44–54, 57–59, 63–68, 97, 207–8, 225, 231n66, 233nn12–13, 235n27; brown commons as, 178; and the Duwamish River, 171, *171*–73, 176, 203
apple trees, 50, 91, 122
aqueducts, 45, 97
aquifers, 12, 30–31, 38, 42–43, *43*, 49, 72, 104, 113, 141, 152; and collective governance, 168; and irrigation, 40, 58; and the Los Angeles River, 5–6, 67; overdrafting of, 67–68, 72, 234n15; recharging, 6, 39, 46, 54, 59–61, 65–66, 69, 71, 130, 220; and underflows, 8; and wells, 50, 56–57. *See also* groundwater; springs; wells
Arctic, 25, 42
Austin, J. L., 246n16
Australia, 142
autohistorias, 83–88, 91–94, 96–97, 99
Aviance, Kevin, 116–19, 130–33
avulsion, 219, 223

bacteria, 12, 18; cyano-, 10, 87
Baird Hatchery, 231n58
bald eagles, 129

273

Bambara, Toni Cade, 248n17
Barad, Karen, 53, 171–72, 202, 231n66, 233n13, 235n27; on apparatus, 32, 44, 46, 50, 59, 176; on intra-action, 32, 44, 64–65
barnacles, 175, 177–79, 184, 200, 203
Bay-Delta Science Conference, 93–94, 100, 211
Beale, John, 197–98, 201
bears, 116, 186, 216
beaver believer movement, 75, 104, 118, 127–28, 244n48
beaver dam analogues, xiv–xv, 40, 45, 71, 76–77, 113, 139
beavers (*Castor canadensis*), xvi–xvii 8, 47–48, 59, 104, 214, 225, 242n8, 243n24, 246n16; "beaver moment," 61–62; beavery modes of thinking, 99; and collaboration, 29, 33, 106–41, 222; dams, xiv–xv, 38, 40–41, 45–46, 61, 71, 75–77, 221–22, 244n48; as ecosystem engineers, 33, 61–62, 107, 133, 156; and reciprocity, 183; removal of, 51; and restoration, 6, 17, 29, 39, 68, 70–71, 74–77, 224, 238n71, 241n1, 244n48; trapping of, 33, 61, 76–77, 107, 110–11, 122–26, 137–39, 220, 243n33, 244n39. *See also* beaver believer movement; beaver dam analogues; State of the Beaver conference
Beaver Valley, 61, 76, 137
"becoming with," 158
bees, 113, 160–61, 242n16
Bellacasa, Maria Puig de la, 139
beneficial use doctrine, xv, 32, 62–63, 69–71, 227n2
Bennett, Harold, 65
Bennett, Jane, 184
Berkeley, CA, 96
Berkshire-Hathaway, 251n6
Bersani, Leo, 154
biodiversity, 34, 86, 133, 136, 155, 233n9

biofilms, 87
biology, 81, 91, 246n16; and beavers, 61, 107, 114–15, 122, 124–27, 134–35, 244n39, 244n48; conservation, 32; evolutionary, 194; field, 22, 154, 164; and salmon, xiii–xxii, 51–52, 111, 114, 242n11
Bird Rose, Deborah, 166
birds, xvi–xvii, 10, 19, 84, 159, 169, 188–90, 193, 201, 205, 219. *See also individual species*
blackberry bushes, xv, 121, 207, 215
Blackfoot Canyon, 79
Black Lives Matter, 100, 194
Black Mesa, 6
Black Power movement, 192
Black radical tradition, 182, 214
Black River, 196
Blanchfield, Brian, 170, 246n2
bobcats, 55
Bodega Water Company, 50–52, 54–55, 236n41
Boeing, 203
Bogan, Mike, 86, 93
Boldt Decision, 187–88
Bourdieu, Pierre, 158, 161, 246n16
Bourque, Shawn, 218, 251n4
Brainard, Jose, 170
Brink, Kenneth "Binks," xiii–xiv
brown commons, 20, 34, 174, 177–205, 227n1, 240n7, 250n62
brown feeling, 80–82
brownfield, 183, 202, 204
Brown Power movement, 192
bulrushes (Cyperaceae), xv, 189
Bustamante, Nao: *Neapolitan*, 88, 92, 96
Butler, Judith, 84, 156
butterflies, 84, 91, 154
Byrd, Jodi, 146–47, 211, 213

California, 159; beavers in, 111, 114–15, 127, 138; Central Coast, 10, 51; climate, 38, 219; Coast Ranges, 115;

droughts in, 107, 147; Indigenous Peoples in, 73–74, 208; levees in, xv; Northern, xiii; salmon in, xviii, 84, 92–93, 99, 111; water diversion to, 45, 48–49; water governance in, 32, 36, 41, 44, 57, 62, 72, 227n2, 237n56. *See also* California Department of Fish and Wildlife (CDFW); California Fish and Game Code; California Regional Water Quality Control Board; California Water Plan Update (2005); *individual cities*

California Department of Fish and Wildlife (CDFW), 41, 115, 122, 138, 233n3, 234n25; *California Fish and Game,* 242n17

California Fish and Game Code, 41, 232n2

California Regional Water Quality Control Board, 234n25, 237n56; Groundwater Study Plan (2008), 65, 69–70; TMDL Action Plan, 70, 232n1, 237n56

California Water Plan Update (2005), 49

Camp Creek, 216

Canada, 16, 42, 215

Cantor, Sierra, 114, 228n17

capitalism, xvii, 47, 90, 147, 183, 196, 198, 202, 233n12, 239n76; anti-capitalism, 144, 197; capitalist ruins, 138; late, 18, 99; liberal, 213

captive breeding, 51–52, 236n39

carbon, 12–13, 18, 30, 36, 88, 90, 125, 184, 219

carbon dioxide, xiii, 163

Carlson, Stephanie, 228n17

Castor canadensis. See beavers (*Castor canadensis*)

Catasetum orchids, 113

cattails, xiv

cattle, 10, 46, 50, 54, 58–60, 63, 74, 77, 85, 111

cedar, 129

Central America, 84

Central Valley, 92

Chambers-Letson, Joshua, 99

chaparral, 243n31

Chen, Mel, 191, 244n50

Chinook salmon. *See* salmon: Chinook (*Oncorhynchus tshawytscha*)

Choy, Tim, 240n14

ciénegas, 6

cisheteronromativity, 14, 20, 22, 32, 126

clams, 19, 175, 184, 187, 209

Clark Fork River, 78

Clayton (fisheries technician), xiv

Clean Water Act, 72, 232n1, 237n56

climate change, 4–5, 38, 46, 62, 86, 90, 100, 115, 158, 216; effect on salmon, 13, 38–39, 59, 71, 111, 113, 124, 214

climate justice, 21, 34, 192

Cluer, Brian, 114

coho salmon. *See* salmon: coho (*Oncorhynchus kitsuch*)

Cole, July Oskar, 106. *See also* Hazard, July

collaborative science, 13, 83

collective action, 5, 15, 128

collective capacity, 15, 220

collective continuance, 15, 155, 187, 197

collective governance, 24, 168

colonialism. *See* settler colonialism

Colorado River, 45

Columbia River, 20, 45, 47, 121, 124, 234n18

comanagement, xvii, 24, 72–73, 140, 188, 236n47

commons, 167, 209, 247n5; brown, 20, 34, 174, 177–205, 227n1, 240n7, 250n62; more-than-human, 32–33, 76; multispecies, 45–50, 54–58, 64, 72, 96, 137–38, 237n55; under-commons, 33, 99, 102, 106, 136, 179, 224

cone of depression, 43
Coney Island, 4
Confederated Indians of Graton Racheria, 239n77
Confederated Salish-Kootenai Tribes, 251n6
conferences (academic/practitioner), 11, 13, 29, 61, 145, 182, 205–6, 229n23, 230n44, 243n34; and affect, 81–85, 153–54; beaver management, 33; performances at, 89, 91–95, 99–100, 211–12; and scientific writing, 98. *See also* American Geophysical Meeting; American Studies Association; Bay-Delta Science Conference; Salmonid Restoration Federation conference; Society for the Social Studies of Science (4S) conference; State of the Beaver conference; *Tell a Salmon Your Troubles*; We Who Feel Differently conference
conservation biology, 32
copepods, 177
co-research, 142, 151, 205
Cosumnes River, 93
cottonwood trees, xvi, 64, 218
Coulthard, Glen, 46–47, 73, 196–98, 210, 215, 237n55, 239n76, 250n49
Courtney, Sofi, 217–18
COVID-19 pandemic, 3, 25, 100, 174, 176–77, 222
coyotes, 55, 115
crabs, 190, 201, 205
Crimp, Douglas, 155
critical development studies, 26
critical physical geography (CPG), 29, 213, 234n14
cruising, 22, 34, 80–81; as methodology, 29, 178, 191, 194, 247n4; and waterfronts, 175, 177–78, 180, 181–82, 185, 188, 191, 202–4
cyborgs, 45, 91, 208

Daigle, Michelle, 29
Dakota People, 15
dams, xvi, 20, 63, 130–32, 213, 225, 233n11, 251n8; beaver/beaver analogue, xiv–xv, 38, 40–41, 45–46, 61, 71, 75–77, 108, 111–27, 135–39, 221–22, 244n48; effect on salmon, xviii, 41, 51, 66, 107, 109, 220, 236nn47–48; removal of, xviii, 16, 20–21, 42, 62, 74, 78–79, 102, 220–21, 228n16; as settler colonial projects, 6, 19, 48, 103. *See also* hydropower
Danger, Annie, 100
Darling, H. Wheels, xiii, xv
Darwin, Charles, 113, 160–61, 242n16
Dave (research collaborator), 167
Dead Horse Bay, 191
death, 12, 17, 55, 108, 227n1; double death, 86, 166, 240n13; mourning of, 85–88, 91, 139, 144, 146–68; premature, 4, 94, 99–100, 112, 153, 158, 164; of salmon, 39, 51, 61, 85–88, 147–53, 159–68, 221. *See also* extinction
decoloniality, 29, 31, 83, 104, 138, 187, 194, 219; and commons, 47, 197; decolonizing ethics, 107, 109, 111, 198; feminist, 80
deer, xiv, 55, 63, 77
Delany, Samuel R., 25–26, 179–80, 247n4
Dellecave, J., 119; *Nocturnal Beaver*, 120, 243n24
deltas, xv, 42, 175
depressive position, 80, 82, 91–92, 95–96
Derrida, Jacques, 246n16
Descartes, René, 44, 176
deserts, 40, 154
diatoms, xv
diffraction, 30, 139, 170–73
Dillon, Lindsey, 250n62

dinoflagellate, 67
disability, 9, 11, 23, 89, 142, 158; crip time, 81–82
dispossession, 73–74, 167, 212, 215–16, 239n76; in *The Gold Fish, or, Straight Flushes for the Manifestly Destined*, 101, 119, 133, 211
dissident feeling, 124, 127
disturbance theory, 248n23
Dohrn, Charlotte: *Substrate*, 189–90
Dolman, Brock, 242n17
downwelling, 38
dredges, xv–xvi, 40–41, 59–60, 126, 197–98, 223
droughts, 7, 16, 28, 38, 74, 107, 127; effect on salmon, 87–88, 92, 111–13, 116, 147, 149, 151, 159–62, 165, 167; and groundwater, 41, 49, 67, 71; and hyporheic zone, 36; at Salmon Creek, 10, 13, 39, 51, 58, 111–16, 147, 159–62, 235n36; water conversation during, 27, 50–51, 53
Duwamish River, 7, 13, 19, 170, 174–78, 180, 182, 187–93, 196–205, 223, 250n55
Duwamish River Cleanup Coalition, 188, 197–98, 201, 203
Duwamish Tribe, 176, 188, 197, 201, 203, 205
Duwamish Waterway Park, 200
dynamic equilibria, 248n23

ecology *vs.* ecologies, 4
ecopoetics, xviii, 26, 30, 34, 169–73, 175–76, 189, 211, 221
Edelman, Elijah Adiv, 17–18
Eel River, 87, 162
electrofishing, 162–63
elk, 63, 77, 244n56
Ellensburg, WA, 122
Elliott Bay, 191
El Niño, 54

emergent properties of ecosystems, 248n23
Endangered Species Act (US, 1973), 51, 62, 242n11
entomology, 81, 86
Environmental Coalition of South Seattle, 250n56
environmental imaginaries, 42, 106
environmental justice, 6, 35, 140, 189, 201; and brown commons, 178, 180, 183, 186, 191–92, 205; and Indigeneity, 22, 27, 90–91, 102, 188; queer, 3–4, 34, 147, 230n43; and restoration, 23; role of science in, 26, 90–91
Erna (research collaborator), 167
erosion, 76, 90, 172, 225
ethics, 14, 31, 35, 44, 80, 85, 131, 145, 218, 224; beaver, 109, 123; decolonizing, 107, 109, 111, 198; ecological, 102; feminist care, 139; and grounded normativity, 210, 239n76; plantation science, 123; queer-trans, 139; of reciprocity, 26, 121; research, 16–17, 22, 29, 162–64; settler, 42, 66, 107, 121, 123, 125; trans, 124, 139
ethnography, 6, 26, 32, 47, 95, 114, 217, 233n11; auto-, 98; kitchen table, 28–29
ethology, 160
Etna City Council, 76
Europe, xvi, 47, 63, 102, 212
evapotranspiration, 53, 112
evergreen huckleberry bushes (*Vaccinium ovatum*), 129
excess, 5, 9, 81, 122, 125–26, 154; queer, 102, 157, 185–86, 191, 204
extinction, 5, 41, 82, 183, 214, 220, 240n13, 241n22; of beavers, 121, 137–38, 235n37; and habitat destruction, 41; mourning, 29, 33, 84–88, 146–68; and queerness, 100;

extinction (*continued*)
   of salmon, 44, 46, 49, 51, 53, 66, 69–70, 90, 93, 111, 116
extraction, xvii, 19, 98, 103, 106–7, 156–60, 189, 233n3; effect on beavers, 125; effect on salmon, xiv; of Indigenous research, 142; resistance to, 7, 192, 198; and settler colonialism, 42, 45, 102, 146, 155, 164, 220

Fabre, Jean Henri, 160–61
farming, 16, 50, 90, 104, 107, 174, 219; and beavers, 125–27, 129; farmer science, 24, 59; and floods, 60–61; and groundwater, 41, 49, 54, 66, 69–70, 74, 103, 113; and habitat destruction, 41, 46; and irrigation, xvi, 72; and queerness, 22; and salmon, 69, 92–93, 234n15
Fawcett, Michael, 51, 85, 111, 164, 228n17
Fay Creek, 85, 111–12, 114, 150–51
Federal Energy Regulatory Committee, 220
Federated Indians of Graton Rancheria, 73, 112, 239n77
femininity, 31, 132, 193
feminist science and technology studies (STS), 13, 27, 31, 48, 213
feminist standpoint theory, 29, 48
Ferguson, Roderick, 168, 182–83, 192
ferns, 129
field praxis, 27, 32–33, 80–102
fig trees, 55
First Nations, 24
fir trees, xiv, 130
Fisher, Frank W., 231n58
fish surveys, xiii, xvii–xviii, 26, 51, 74, 87, 138, 162–64
Fish Wars, 187
floods/floodplains, 5, 8, 10, 16, 38, 59, 74, 109, 120, 140–41, 150, 204, 223, 225; as amphibious, xvi, 222; and beavers, 40, 61, 75, 77, 108, 112–14, 123–27, 129, 136–37, 238n71; dams producing, 103, 107; and El Niño, 54; images of, *43*, *108*; Indigenous *vs.* settler conceptions of, xvii, 7, 15, 20, 24, 125–26, 218–20; and irrigation, xvi, 60; mega-, xv; and mining, xvi, 40, 60, 132, 136, 215; restoration/reconnection projects, xvii, 6, 39, 68, 70–71, 113–14, 219, 238n71; and salmon, xvii, 64, 159, 251n6; and waterfronts, 176–77, 179, 181, 191, 196, 198; and water governance, 49, 63
flowers, 57, 81, 160–61, 172, 216
forest plantations, 124, 174
forestry, 22, 32, 39, 125. *See also* US Forest Service
forests, 53, 73, 84, 112–13, 116, 139, 174; fires in, 25, 48, 59, 107, 208, 218, 244n48; management of, 73–74, 123, 138, 208; plantations, 124; restoration of, 157; riparian, xvii, 60, 220; second-growth, 56
Frank, Billy, Jr., 23
Freeman, Elizabeth, 33, 156–59, 161, 246n16
French Creek, 59
frogs, 58, 86
fugitivity, 36, 135–36

garlic, 55
Garrard, Greg, 233n8, 233n11
Gay Power movement, 192
geese, 114, 172, 176, 202
gender, 11, 17–18, 22, 44, 89, 133, 198, 228n4, 230n48; and beavers, 128, 130–31; and embodiment, 79, 104; gendered violence, 4, 215–16, 218; genderfuck, 4, 223; genderqueerness, 110, 177; misogyny, 28, 133; and queer kinship, 142–45, 217; and queer performance, 116–18, 121; in science, 17, 21, 28, 31–35, 44, 78–79, 82, 85, 92, 145. *See also*

cisheteronromativity; femininity; feminist science and technology studies (STS); heteropatriarchy; masculinity; transphobia
genomic sciences, 15
geology, 6–7, 32, 50, 78, 90, 135–36, 228n16
geomorphology, 114
Gillespie, Katie, 245n4
Gilmore, Charna, xiv–xv, xviii
Glacier National Park, 130
Glanzberg, Joel, 228n15
Glitter Bomb, 93
GLQ, 179
Goldstein, Alyosha, 211, 213
Gordon, Avery, 182, 248n17
Gould, Deborah, 155
Gowanus Canal, 191
grapevines, 50, 54
Green New Deal, 204
grief, 4, 34, 127, 144, 146, 205, 245n4; for species extinction, 5, 29, 33, 84–88, 92–100, 138, 146–68. *See also* melancholia; mourning
grounded normativity, 73, 210, 239n76
groundwater, xiii, xviii, 7–8, 13, 79, 125, 149, 236n47, 237n56; and beavers, 75–77, 108, 224, 244n48; governance of, 36, 41, 49, 57–59, 65–72; and irrigation, 60, 66, 72, 236n49; recharging of, 27, 37, 43, 46, 66–71, 76, 108, 165, 244n48; relation to surface water, 36–39, 43, 43–44, 46, 49, 54, 56–58, 63, 66–75, 103, 132, 168, 223–24; and salmon, 11–12, 60, 62–64, 66–70, 88, 113, 151, 152. *See also* aquifers; springs; wells
groundwater studies, 56, 62, 65, 69–70, 168
Groundwater Sustainability Agency, 41
Group f/64, 208
Guadalupe Creek, 18

Guatemala, 84
gulls, 207

habitus, 33, 158–59, 161, 246n16
Halberstam, Jack, 194
Hall, Stuart, 182
Hamm Creek, 201
Haraway, Donna, 158
Harbor Island, 176
Harding, Vincent, 182
Harling, Will, 130, 134–37, 244n48
Harney, Stefano, 33, 106, 136, 179
Harvey, Jasmine, 251n4
hatcheries, 48, 51–52, 98, 111, 124, 231n58, 236n39
hay, 40, 59, 63
Hayward, Eva, 104–6, 109–10, 137
Hazard, July, xv, 93, 98, 104–6, 110, 129–30, 133, 135, 178, 202–3, 246n2; and ecopoetics, 30, 169–70, 189, 246n1; *The Gold Fish, or, Straight Flushes for the Manifestly Destined*, 100–101, *101*, 119–20, 211; and queer ecologies, 172, 174–76; and queer kinship, 144; work with beavers, 107, 122, 124–25. *See also* Queer Ecologies tour; *Tell a Salmon Your Troubles*
heʔapus, 197
Hena, Louie, 228n15
Herko, Fred, 99, 194
Herring's House, 201
heteropatriarchy, 130–31, 147, 239n76
heterosexism, 83, 156, 223. *See also* homophobia
Hillman, Leaf, 34, 214–15, 217
Hitchcock, Alfred: *The Birds*, 50
HIV/AIDS, 4, 154–55. *See also* ACT UP
Holling, C. S., 248n23
Homo Cult, 178
homonormativity, 89, 131
homophobia, 28, 117, 155. *See also* heterosexism
Hong Kong, 240n14

Hoopa Tribe, 59, 236n47
hopeless hopefulness, 90, 100, 168
humanities, 98, 143, 147, 175, 211; geohumanities, 42. *See also individual disciplines*
hunting, xiv, 197
Hurricane Sandy, 120
Hustak, Carla, 160–61, 241n3
hydrology, xviii, 8, 12, 26, 35, 104, 109, 218, 224; and beavers, 61, 124–25, 132, 138; and connectivity, 46; eco-, 32, 36, 97; and hyporheic zone, 36, 38, 43; Indigenous, xvii, 6; and salmon, 67–68, 74, 88, 93, 148, 152, 234n15; and water governance, 45, 48–50
hydropower, xv, 19. *See also* dams
hyporheic connection, 32, 45, 48–49, 54, 56, 58
hyporheic zone, xviii, 8, 36–39, 43, 81, 222

Ich (*Ichthyophthirius mulifiliis*), 62
indeterminacy, xvi, 25, 67, 124, 225, 237n64
Indigenous law, 44; treaty law/rights, xvii, 15, 17, 24, 42, 63–66, 68, 73, 188, 197, 205, 236n47
Indigenous protocol, 6, 15–17, 26, 73, 213–14; Karuk, xviii, 134, 215–17, 221; Quartz Valley Indian Community, 42
Indigenous research methodologies, 13, 16, 26, 29, 142–43, 145
Indigenous science, 6, 15, 17, 24, 48, 75, 102, 121, 142, 166
Indigenous studies, 6, 16, 29, 80, 213–14
inflow, 36, 97
infrastructure, 15, 63–65, 71, 104, 193, 220, 238n71, 248n22; and beavers, 59–61, 77, 127; and household water use, 44–46, 53, 58; shaping river flow, xv–xvi, 7; and waterfronts, 174–77
insects, xv, 37, 58, 84, 95, 160–61, 164, 205, 219
Integrated Regional Watershed Management Plans, 49
intimacy, 33–34, 57, 84, 86, 93, 95, 150, 160, 187, 239n76; intimate violence, 4
intra-action, 32, 44, 58–59, 64–65, 176, 184, 225
involuntary momentum, 160
iron, 13
irrigation, xv–xvi, 40–41, 45, 52, 54–55, 58, 77, 103; ditches, xvi, 39, 59–60, 62–63, 68, 72, 76, 109, 125, 129; and groundwater, 49–50, 58–69, 76; and settler colonialism, xiv, 19

Japan, 124
Jasanoff, Sheila, 233n12
Jay (creek-walker), 114
Jennifer (biologist), xiii–xiv
Jones, Bill T., 117
Jordan, Frank C., 137
joy, 88, 92, 98, 105, 116, 138–39, 149, 156, 159, 167–68, 203
Just, Tony, 119

Kalispel Tribe, 130
Karuk Department of Natural Resources, x, 27, 134, 244n56, 251n4
Karuk Fisheries, xiii–xiv, 70, 76, 244n56
Karuk Tribe, xvi, 32, 34, 59, 66, 68, 74, 112, 130, 134, 138–39, 218, 222–23, 234n15; fishing rights, 236n47; protocol, xviii, 134, 214–17, 221. *See also* Karuk Department of Natural Resources; Karuk Fisheries
Kauanui, J. Kēhaulani, 74
Kelson, Suzanne, 86–87, 162–64

Kerridge, Richard, 233n8, 233n11
Kim, Sang-Hyun, 233n12
kinship, 6, 31, 135–36, 148, 245n4; practical, 147, 155–58, 161, 165, 167; queer, 13–15, 17, 21, 24, 28, 138–39, 142–45, 147, 154–58, 167, 177, 223; transspecies, 33, 94, 115, 126, 147, 155–59, 165, 246n16; white, 24
Klamath Basin, xv, xvii, 13, 42, 137, 236n48
Klamath Basin Tribal Water Quality Work Group, 66, 234n25
Klamath River, xiv, xviii, 7, 27, 34, 60, 62, 110, 244n48; beavers in, 110, 130, 134, 137; salmon in, xvii, 62, 112; tribes, xvii, 60, 112, 130, 215
Klamath Riverkeeper, 66
Klamath Valley, 110
Kootenai Tribe, 130. *See also* Confederated Salish-Kootenai Tribes
Kwakwaka'wakw People, 129

Lake, Frank, 218
Lake Washington, 172
latency, 8, 14, 20, 103–5, 140, 218
Lau, Sallie, 170
legacy impacts, 60, 63, 70. *See also* dredges; levees; mining
Les Misérables, 100, 211
levees, xv, 20, 103, 109, 129, 132, 191–92, 196, 219; effect on salmon, 51, 60–61, 63–64, 66, 70, 111, 234n15; and habitat destruction, 41; and settler colonialism, 48, 220, 223
Lin, Maya, 241n22
Linton, Jamie, 20, 45, 233n11
Linzy, Kalip, 89
Living Structures, Inc., 228n15
locust trees, 215
logging, 22, 51, 57, 84, 112, 138
Lopez, Patricia, 245n4
Los Angeles, CA, 78, 186, 193, 243n31

Los Angeles Metropolitan Water District, 62. *See also* Mono Lake Decision
Los Angeles River, 5
Love Canal, 192
Lundquist, Kate, 114–16, 124, 138, 242n17

MacArthur Park, 195
magnesium, 13, 54
Manifest Destiny, 126, 129, 136, 139, 141, 176, 208, 211–14; and beavers, 109, 122, 138; dams/levees in, 48, 103; effect on salmon, xv; resistance to, 8, 14, 17, 20, 101–2, 104, 107, 131, 186, 192
"The Manifest Reversals of Multispecies Watershed Restoration," 140–41
manzanita bushes, 28, 216
March for Black Trans Lives, 3
marshlands, xvi–xviii, 175–76, 181, 191, 198, 199
Martinez, CA, 127
masculinity, 126, 132, 217; toxic, 31; transmasculinity, 28, 133
Masura, Diane, 56057
Matzen, Qilo, 101
Mauss, Marcel, 161
McCloud River, 231n58
McCovey, Kathy, 216
meadows, 39, 41, 59, 99, 107, 122, 131
Mediterranean climates, 38
Meek, Stephen, 61, 137
Melamed, Jodi, 211, 213
melancholia, 82, 161, 225. *See also* grief; mourning
mercury, 18–19, 102, 184, 199, 249n32, 250n55
Mermaid Parade, 4
methodology of book, 5–14, 25–32, 47, 146–47, 158, 168, 233n11, 235n36

Methow Forest Service, 127
Methow River, 107, 122, 124, 138
Methow Valley, 107, 110, 124, 128–29, 236n53
Métis People, 15, 184
Mexico, 84
microbes, 13, 36–37, 37, 47, 82, 87, 91, 157, 184, 190, 201, 219
Mid-Klamath Watershed Council, xvii, 27, 244n56
Mill Creek, 242n8
Milltown Dam, 78, 228n16
mimicry, xiv, 68, 115, 118, 160–61, 224
mining, 39, 41, 63–66, 126, 136, 217; and gender violence, 215, 218; tailings, xiii–xiv, xvi, 40, 46, 59–60, 132, 215–16, 218, 223, 238n71
Miwok Tribe, 73, 112. *See also* Federated Indians of Graton Rancheria
MIX NYC, 120
Modoc Tribe, 208
Mohawk Nation, 216
Mono Lake Decision, 62
Montana, 7, 78, 228n16, 246n2
Monterey Bay, 178
Moore, Johnnie, 228n16
Morehead-Hillman, Lisa, xviii, 27, 34, 214–15
Morris, Mark, 117
Morrison, Leah, *101*
moss, 129, 175
Moten, Fred, 33, 106, 136, 179
mourning, 17, 34, 144, 154; multispecies, 29, 33–34, 84–88, 92, 155, 158, 164. *See also* grief; melancholia
Muckleshoot Tribe, 182, 187–88, 197, 201, 205
multispecies commons, 45–50, 54–58, 64, 72, 96, 137–38, 237n55
Muñoz, José Esteban, 3, 32, 83, 85, 130, 138, 177, 190, 202–6, 211, 248n22, 249n36; on brown commons, 20, 174, 178–87, 192–93, 196–98, 227n1, 240n7, 250n62; on Bustamante, 95–96; on critical utopianism, 250n53; on cruising methodologies, 29, 194, 247n4; on the depressive position, 80, 82, 91–92, 96; on excess, 125; on Fred Herko, 99; on hopeless hopefulness, 90, 168; on Jack Spicer, 185–86, 247n5; on Kevin Aviance, 116–17, 119–21, 131–33; premature death of, 100, 178, 247n2; on queer ecology, 34, 193, 195, 249n42; on the trace, 29, 118, 129, 131–33, 201; on waiting, 82, 89; on white affect, 249n41, 250n62
Murphy, Michelle, 155, 167
mushrooms, 82, 123, 157
mussels, 19, 170, 176, 199
Myers, Natasha, 160–61, 241n3

Nancy, Jean-Luc, 195
National Marine Fisheries Service, 51, 220, 234n25
National Oceanic and Atmospheric Administration (NOAA), 61, 114
natural science, 6, 21, 133, 143, 145, 179. *See also individual disciplines*
Navajo Nation, 6
Navarete, José, *101*
Neimanis, Astrida, 233n8, 233n11
neoliberalism, 90–91, 98
Nepon, Ezra Berkley, 119; *No One Mourns the Wicked,* 120
New Almaden, CA, 249n32
New York City, 29, 83, 116–18, 120, 130, 178, 194, 240n7, 243n23; Brooklyn, 3, 191
Nisqually Tribe, 251n6
nitrogen, 36, 88, 116
nongovernmental organizations (NGOs), 42, 44, 60, 62, 66, 205, 229n26, 233n9
North America, 47, 84, 142

North Coast Regional Water Quality Control Board, 65, 232n1
Nyong'o, Tavia, 82, 89, 194

Oakland, CA, xv, 177; waterfront, 19, 179, 191, 199
objectivity, 16–17, 29, 48, 154, 162, 164, 234n16
olive trees, 55
onions, 55
ontologies, 17, 67, 118, 140, 142, 224, 237n64
orchards, 50, 122, 124–25
Oregon, 61, 75, 124, 209
Orleans Valley, 135
other-than-human beings, 8–10, 14, 18–19, 21, 116, 134, 164, 197
otters, 52
outflow, 36, 97
owls, 115
oxygen, 36–37, 37, 113, 148, 151–52, 159, 184; loss of, 10, 12–13, 52, 84, 87–88, 112, 149, 152
oysters, 102

Pacific lamprey (*Entosphenus tridentatus*), 60
Pacific Northwest, 33, 102, 107, 111, 251n6
PacifiCorp, 251n6
Pacific salmonids, xiv
Palestine, 143, 147
Parreñas, Juno, 241n18
partial perspectives, 16, 234n16
participatory action research, 26
Paulatuuq People, 42
PCBs (polychlorinated biphenyls), 19, 184, 191, 199–200, 204, 250n55
performance, 7, 29, 110, 128, 192, 200, 208, 225, 246n16; endurance, 12, 120, 243n24; and mourning, 33–34, 96, 154; queer, 89–90, 99, 116–21, 131–33, 157, 178–81, 185–86, 193–96, 211–12, 221–23; and science, 81–83, 90–95, 99–102, 115, 119–20, 140–41, 147–49, 167, 189, 221–23; of whiteness, 125–26
performativity, xviii, 89, 113, 119, 179, 181, 186, 204, 225
Pfiesteria, 67, 237n64
pickleweed, 202
Píkyav Field Institute, 216
pikyávish, 215–17
Pine Gulch, 83
pine trees, 130, 154, 157
PIT (passive integrated transponder) tags, xiii, 11, 124, 227n1
plankton, 19, 184, 190
plantation science, 123, 158
police violence, 3–4, 17, 25, 143, 155
Pollock, Michael, 61, 75–76, 127, 135, 139
Pomo Tribe, 73, 112, 114, 239n77. *See also* Federated Indians of Graton Rancheria
Portage Bay, 171, 189
Port of Seattle, 175
Portugal, 55
postcolonial studies, 16, 109, 218
potassium, 30
potatoes, 55, 216, 218
Potawatomi Nation, 15, 102
Power, Mary, 228n17
prairies, 73
private property, xiv, 85, 132, 196–98; and beavers, 122, 125; and groundwater, 68; and settler colonialism, 42, 47, 109
Puar, Jasbir, 34
public trust doctrine, xv, 62–64, 66, 68–70, 227n2
Purim, 120

Quartz Valley Indian Community, xvi–xviii, 27, 59–60, 238n65;

Quartz Valley Indian Community (*continued*)
    water governance work, 41–42, 47, 65–72, 74, 237n63
Quartz Valley Indian Reservation (QVIR), 60, 236n47
queer and trans of color critique, 3–4, 80, 182–83, 187, 213
Queer Ecologies as Environmental Justice Strategy, 230n43
Queer Ecologies tour, 170, 174–76
Queer Nation, 100
queer science, definition, 4
queer theory, 5, 105, 156, 179, 244n46. *See also* queer and trans of color critique
queer trans, definition, 4
Queer Utopias event, 179–80
queer x trans x feminist ecology, 31–32, 80–102

raccoons, 53, 55, 87–88, 116, 159, 178, 186
race, 100, 142–43, 145, 165, 168–69, 192, 205, 214, 223, 250n62; and affect, 249n41; and animacy, 136, 191, 244n50; and beavers, 120, 127; and COVID-19, 177; and death, 3, 17–18, 147; and embodiment, 83; and environmentalism, 208; and kinship, 24, 167; performance of, 126; and police violence, 25; and queer land projects, 22, 230n51; and queer performance, 89, 116–18, 131–32; racial purity logics, 74; in ranching, 59; in science, 11, 21–23, 26, 31–35, 44, 73, 82–83, 89, 154; and settler colonialism, 3–5, 9, 15, 29–35, 48, 76, 107, 147, 167, 181, 203–4, 210–13, 228n4, 239n76, 251n5; and waterfronts, 19, 177, 179–83, 187, 194–98, 201. *See also* antiracism; brown commons; brown feeling; queer and trans of color critique; racial justice; racism; whiteness; white supremacy
racial justice, 21
racism, 19, 28, 32, 74, 82–83, 145, 158, 181, 187; anti-Black, 22, 25, 29, 117, 210, 228n4. *See also* antiracism; police violence; white supremacy
Radical Faeries, 22
rain tanks, 12, 50–55, 113
Ramirez, Margaret Marietta, 29
rancherias, 73, 112, 239n77. *See also* reservations
ranching, xiii, 45, 50–51, 55, 112–13, 122, 132, 139, 234n15, 238n65; and beavers, 46, 61–62, 75–77, 114, 125–27, 138; and cisheteronormativity, 22; conservation, 74; water use, xvi, 41–42, 54, 59–65, 67–71
Rasmussen, James, 197
reciprocity, 15, 35, 42, 44, 72, 85, 110, 126, 134, 183–84, 218; ethics of, 26, 121; and grounded normativity, 210, 239n76; and Indigenous research methodologies, xvi, 13, 27, 143, 168
reconciliation ecology, 23
recovery, 15, 110, 196, 205, 213, 234n15, 235n37, 236n48, 251n6; beaver, 6, 137; definition, 23; salmon, 6, 27, 39, 51–52, 58, 67–71, 77, 111, 113, 204
Reddy, Chandan, 211, 213
Red Power movement, 192
redwood trees, 50, 56, 151
rehabilitation, 6, 23, 143, 223
relationality, 7, 20–21, 128, 160–61, 171, 189–90, 213, 222–23; and beavers, 138; grounded, 34, 180, 210, 214; oil, 184; and queerness, 34, 109, 117, 119, 143, 147, 158–59, 165–68; in research methodologies, 13, 24, 27–28, 143, 146–47, 157–58, 165–68, 180, 205; and settler colonialism, 126, 239n76

reparation, 23, 91, 183, 194, 196
reservations, 41, 66, 73, 236n47. *See also* Quartz Valley Indian Reservation (QVIR); rancherias
reservoirs, 12, 38, 78–79, 221
restoration, 5, 24, 107, 140, 176–77, 181, 191, 207, 220, 228n15; beavers assisting, 17, 29, 33, 39, 61, 68, 70–71, 74–77, 104, 107–18, 122–29, 132, 134–38, 222, 241n1; of chaparral, 243n31; definition, 23; ecocultural, xviii, 6, 27, 72–73, 134; of floodplains, 68, 114; of forests, 157; of headwater meadows, 59; of high-mountain meadows, 39; of rivers, 16, 20, 30, 32, 33, 47, 118–19, 121–23, 195, 213, 215, 229n26, 242n11; of salmon habitats, 17, 28–29, 60–64, 69–70, 109–14, 132–34, 193, 244n48; of streams, 7, 22; of tidal marshes, 175; of urban waterfronts, 191–205; and water use, 53
resurgence, xviii, 21, 23, 180, 225; Indigenous, 24, 72, 107, 109, 201, 215, 239n76
Richmond, CA, 19
riparian forests, xvii, 60, 108, 220
riprap, 11, 140, 175, 189, 192, 199, 201, 207
river model of justice, xviii, 32, 34, 221
river science, 11, 21, 26–27, 35, 38, 45, 109; queer trans feminist, 3–4, 9, 23, 212, 214, 221
Rohlfs, Rori, *101*
rushes (Juncaceae), xiv
Russia, 114
Russian River, 50–52, 111, 124, 138, 236n39, 242n11
Russian River Captive Broodstock Program, 52, 236n39

Sacramento, CA, 93
Sacramento River, 7, 45
Sacramento–San Joaquin Delta, xv
salamanders, 58
salmon: Chinook (*Oncorhynchus tshawytscha*), xiv, 41, 60, 62, 70, 199, 231n58
salmon: coho (*Oncorhynchus kitsuch*), 13, 85–86, 112, 162, 231n63, 235n37; in Fay Creek, 114, 150–51; in Salmon Creek, 39, 46, 51–52, 58, 62, 68, 111, 114, 147, 159, 242n11; in Scott Valley, xiv–xvii, 41, 60, 70, 236nn47–48
Salmon Creek, 18, 27–28, 39, 44, 232n2, 235n36; beavers at, 110–16, 138; compared to Scott Valley, 62–63; drought at, 10, 13, 147; and multispecies commons, 46–47, 50–58; salmon in, 13, 39, 44, 48–59, 86–88, 153, 158–63, 242n11; and settler colonialism, 73–74; water governance of, 32, 132, 168
Salmon Creek Watershed Council, 51, 56–57, 113, 165–66, 168, 234n25
salmon death, 39, 51, 61, 85–88, 147–53, 159–68, 221
salmon extinction, 44, 46, 49, 51, 53, 66, 69–70, 90, 93, 111, 116
Salmonid Restoration Federation conference, 243n34
Samuels, Ellen, 81–82
San Fernando Valley, 5
San Francisco, CA, 19, 50, 55, 93, 177, 196; Embarcadero, 191; Hunters Point, 199
San Francisco Bay, 18, 178, 249n32
San Francisco Bay Area, 6, 198
San Joaquin Valley, xv, 196
San Jose, CA, 18
Sarna-Wojcicki, Daniel, xv, 218, 227n3, 232n2, 234n15, 243n34; work with beavers, 61, 74, 140, 221–22, 238n71, 244n48. *See also* "The Manifest Reversals of Multispecies Watershed Restoration"
schema of resilience, 248n23

Schrader, Astrid, 67, 237n64
Schuller, Kyla, 146, 154, 167
science and technology studies (STS), 14, 24, 30, 94; feminist, 13, 27, 31, 48, 213
science communication, 34, 86
Scottish Power, 251n6
Scott River, xiv, xviii, 40, 46–48, 232n1; beavers in, 76; and groundwater, 64–67; and irrigation, 45, 59; salmon in, 60, 70, 237n56
Scott River Watershed Council (SRWC), xiv, xvi–xviii, 27–28, 60, 65, 75, 234n25; Quartz Valley Indian Community collaboration, 71, 74; work with beavers, 40–41, 76–77
Scott River Water Trust, 67–69, 71–72, 237n63
Scott Valley, 28, 47, 139, 223–24, 232n2, 236n53, 238n65; beavers in, 59, 61–62, 68, 71, 74–77, 137; salmon in, xiii–xvii, 44, 62–72, 128–29, 137, 236nn47–48; water governance in, 32, 43, 49, 58–73; water use in, 41–42
Scott Valley Groundwater Advisory Council, 234n25
scour, 61, 79, 109, 212, 218–19, 225
Seafarers Rest Mission, 202
Seattle, WA, 5, 174, 175–77, 188, 191, 199, 216–17, 250n56; rivers, 19, 174, 180, 182, 193, 196, 200, 202. *See also* Port of Seattle; West Seattle Bridge
Sedgwick, Eve Kosofsky, 32, 83–84, 154
Seiad Creek, 110, 244n48
Seiad Valley, 128–29
settler colonialism, 83, 94, 104, 112–14, 140, 175–76, 180, 214, 216–17, 221–25, 229n26; and beavers, 61, 76–77, 116, 118, 122–36, 244n48; and commons, 46–47, 63, 167–68; and dams, 6, 19, 48, 103; and enclosure, 8; ethics of, 42, 66, 107, 121, 123, 125; and extraction, 42, 45, 102, 146, 155, 164, 220; and floods/floodplains, xvii, 7, 15, 20, 24, 125–26, 218–20; and Indigenous science, 16–17, 24; industrial, 73, 212; and irrigation, xiv, 19; and levees, 48, 220, 223; and private property, 42, 47, 109; and queerness, 14–15, 20, 22–23, 118–21, 130–31; and race, 3–5, 9, 15, 29–35, 48, 76, 107, 147, 167, 181, 203–4, 210–13, 228n4, 239n76, 251n5; and relationality, 126, 239n76; and Salmon Creek, 50, 73–74, 110, 116; and Scott Valley, xiv–xv, 42, 65–67, 71–72; settler science, 7, 17, 24, 61, 114; and Skagit Valley, 234n15; and stewardship, 42, 63, 155; and waterfronts, 192, 196–98, 203, 205; and water governance, 41–42, 44–45, 58, 63–67, 71–74. *See also* Manifest Destiny; terra nullius
Seward, Marilyn, 76
sex work, 21–22, 191, 204
Shackleford Creek, 41, 60, 66, 72, 236n47, 237n63
Shasta Peoples, xviii
Shasta-Trinity alps, 59
shellfish, 102, 182, 191, 194, 199. *See also individual species*
shrimp, 50
Silicon Valley, 18, 249n32
Silver Platter, 186, 191, 193–94
Simpson, Audra, 216
Simpson, Leanne Betasamosake, 73, 210, 215, 231n64, 239n76
Sioux Peoples, 208
Siskiyou County, CA, 41, 44, 60, 66, 70, 75
Siskiyou County Farm Bureau, 41
Siskiyou County Superior Court, 41
Siskiyou Resource Conservation District (RCD), 60, 66, 70

situated knowledges, 75
Skagit Valley, 234n15
slime molds, 37, 85–86, 191, 194
snails, 86, 208
social science, 6, 8, 20, 30, 94, 98, 133, 179. *See also individual disciplines*
Society for the Social Studies of Science (4S) conference, 92, 94
sodium, 30, 195
SomARTS Gallery, 93
Sonoma County, CA, 44, 165
Sonoma Creek, 242n8
Spain, 114
species loss, 155, 224. *See also* extinction
Spicer, Jack, 185–86, 247n5
springs, 8, 12, 36, 45, 149–52, 154, 168; and Salmon Creek, 13, 50–58, 113, 165
sq̓ʷuʔq̓ʷuʔ (Confluence of Waters), 196
Standing Rock, 143, 192
Stapleton, Betsy, 77
star thistle, 215
State of the Beaver conference, 236n53, 243n34
steelhead trout. *See* trout: steelhead (*Oncorhynchus mykiss*)
stewardship, 53, 55, 197; Indigenous, 64, 102, 218; settler, 42, 63, 155
Stillaguamish River, 7
Stillaguamish Tribe, 251n6
stochasticity, 8, 30, 33, 106, 120–21, 124, 127, 136, 138–39, 193
Stone, Livingstone, 231n58
Stonewall Rebellion, 25, 147
straight science, 6, 31, 167, 189
subjugated knowledges, 4, 16, 44
Subramaniam, Banu, 225, 243n31
Sugar Creek, xiii, 40, 59
Superfund, 20, 180, 188, 195, 197, 199–203
surface water, 13, 20, 237n56; relation to groundwater, 36–39, 43, 43–44, 46, 49, 54, 56–58, 63, 66–75, 103, 132, 168, 223–24
survivance, 107, 175, 237n55
Sustainable Groundwater Management Act (California, 2014), 41, 49, 71–72, 237n57, 238n69
swamps, xvii, 61, 181
Swinomish Tribe, 234n15

T-107 Park/Ha-Ah-Poos (*heʔapus*)—Duwamish Village Park, 188
Tafoya, Terry, 142
TallBear, Kim, 15, 72, 197, 245n4
Tannery Creek, 111, 150–51
Taos/Warm Springs Peoples, 142
Taylor, Ceyenne, 3, 18
Tell a Salmon Your Troubles, 91–95, 99–100
terra nullius, 23
Tesuque Pueblo People, 228n15
"thinking with," 32–33, 40–77, 109, 118, 138, 229n20, 241n3, 242n16; definition, 241n3; as method, 133
Third District Court of Appeal, 41
Tishánik, 214–18, 221, 223
Todd, Zoe, 15, 42, 184
toxicity, 6, 67, 91, 175, 177, 187, 198–200, 209, 221, 250n55; toxic cleanup, 21, 191–92, 203, 205; toxic masculinity/femininity, 31
the trace, 109, 114, 130, 135, 175, 178, 190, 191, 199, 241n3; Muñoz on, 29, 99, 116, 118–21, 131, 133, 179, 196, 201
transing, 105, 133
transphobia, 28, 117
trans sciences, definition, 4
trans theory, 5, 9, 21, 106, 109, 134–35. *See also* queer and trans of color critique
trans vitalities, 17–18
trapping, 33, 61, 76–77, 107, 110–11, 122, 123, 123–25, 124, 127–39, 220, 243n33

INDEX 287

treaty era, 73
treaty law/rights, xvii, 15, 17, 24, 42, 63–66, 68, 73, 188, 197, 205, 236n47. *See also* Boldt Decision
tributaries, xvii, 7
Trotsky, Leon, 206
trout: steelhead (*Oncorhynchus mykiss*), xvii, 13, 39, 61, 85–86, 167, 231n58, 231n63; in Eel River, 162–63; in Fay Creek, 112, 114, 150–51; in Salmon Creek, 68, 87, 111, 114, 147, 159; in Scott Valley, xiv, 41, 60
Tsang, Wu: *Wildness*, 186, 192–95, 249n42
Tsing, Anna, 123, 138, 157–58, 170, 175, 233n9, 249n42
Tuck, Eve, 187–88, 205, 251n5
Tulalip Tribe, 107
Tulare Lake, xv
tuna, 19
turtles, 52
Twitter, 33, 86, 88, 93–94, 142. *See also* #CuteOff

Umpqua River, 7
undercommons, 33, 99, 102, 106, 136, 179, 224
underflows, definition, 8–9
University of California, Berkeley, 145, 244n56; Carlson Lab, 83, 145
University of California, Davis, 62, 65–66, 68–69, 72, 234n15, 238n65
University of California Angelo Reserve, 162
University of Montana, 228n16, 246n2
University of Washington (Seattle), xvii, 74, 188, 216–17, 230n43; Green Futures Lab, 189
upwelling, xvii, 31, 38–39, 47, 61, 224
US Army Corps of Engineers, 5, 60, 204
US Bureau of Reclamation, 45

US Department of Agriculture, 76
US Environmental Protection Agency (EPA), 65, 197, 232n1
US Fish and Wildlife Service, xiii, 244n56
US Fish Commission, 231n58
US Forest Service, xvii, 107, 122, 138, 216, 218, 244n39, 244n56
US Geological Survey (USGS), 67, 69, 238n65

Vay, Sarola Jane: *The Gold Fish Casino*, 243n23

waiting, 82, 88–89, 171, 199, 246n16
Warm Springs conservation hatchery, 111
Washington (state), 107, 187. *See also individual cities*
Washington Department of Fish and Wildlife, 122, 138
waterfronts, 14, 21–22, 30, 32, 34–35, 174–206
Water Protectors, 100, 194, 220
watershed feeling, 124, 126
water table, xv, 30, 36–37, 37, 40, 43, 56, 61, 67, 72, 76, 103, 108, 219
Water Underground, 29, 93, 119
wells, 43, 49–59, 63–69, 103, 113–14, 129, 166, 209; monitoring of, 11–13, 27, 56, 69, 168, 238n65
Western science, definition, 24
West Seattle Bridge, 176, 202
wet-dry mapping, 27, 138, 166
We Who Feel Differently conference, 240n7, 248n22
Wheat, Eli, 230n43. *See also* Queer Ecologies as Environmental Justice Strategy
White, Bernice, 202
White, Richard, 20, 45, 47, 234n18
whiteness, 3, 32, 125–26, 145, 154, 180, 203, 250n62

white supremacy, 15, 117, 147, 167, 187, 210, 239n76; and queer land practices, 22; resistance to, 9, 29, 120, 131, 177, 223, 250n62; in water policy, 48
Whyte, Kyle Powys, 15, 23, 72–73, 102, 212, 220
Wilcox, Andrew, 228n16
Willey, Angie, 225
Willow Creek State Park, 115
willow trees, xiii–xiv, xvi–xvii, 8, 36, 50, 162, 170, 201–2, 215, 218; and beavers, 40, 77, 110, 122, 125–26, 130–31, 135–37, 141, 222
Wilson, Shawn, 142–43, 145
Winnemem Wintu Tribe, 231n58
Wojnarowicz, David, 154
Women's Land, 22
Woodruff, Kent, 127, 135, 244n39
World War I, xv
worms, 161, 208–9

Yakama Nation, 107, 122, 134, 138
Yakima River, 124, 138
Yakima Valley, 126, 236n53
Yang, K. Wayne, 187–88, 205, 251n5
yellow warbler, 84–85, 154
Yokel, Danielle, 76
Yoshiyama, Ronald M., 231n58
Yurok Tribe, 59, 66, 68, 112, 138, 236n47, 244n56

Cleo Wölfle Hazard is an artist-scholar whose work includes photography, video, street theater, and scientific investigation as participatory performance. Originally from Los Angeles, he has worked on restoration crews in New Mexico and in environmental justice education in the San Francisco Bay Area. His scientific collaborations with Native nations and grassroots groups have explored dam removal, salmon and beaver recovery, cultural fire and future streamflow, and industrial contamination, among other phenomena. He coedited the activist anthology *Dam Nation: Dispatches from the Water Underground* (Soft Skull, 2007). Hazard lives in Seattle with his partner, the poet July Hazard, kid Blue Jay, nine chickens, and a cat named Junk Tiger.

Feminist Technosciences
Rebecca Herzig and Banu Subramaniam, *Series Editors*

Figuring the Population Bomb: Gender and Demography in the Mid-Twentieth Century, by Carole R. McCann

Risky Bodies and Techno-Intimacy: Reflections on Sexuality, Media, Science, Finance, by Geeta Patel

Reinventing Hoodia: Peoples, Plants, and Patents in South Africa, by Laura A. Foster

Queer Feminist Science Studies: A Reader, edited by Cyd Cipolla, Kristina Gupta, David A. Rubin, and Angela Willey

Gender before Birth: Sex Selection in a Transnational Context, by Rajani Bhatia

Molecular Feminisms: Biology, Becomings, and Life in the Lab, by Deboleena Roy

Holy Science: The Biopolitics of Hindu Nationalism, by Banu Subramaniam

Bad Dog: Pit Bull Politics and Multispecies Justice, by Harlan Weaver

Underflows: Queer Trans Ecologies and River Justice, by Cleo Wölfle Hazard